Why does a Nucleus Stay Together If Protons (+) Repel Each Other?

A Nucleus is Just . . . a Nucleomagnetics Ring

The Nucleomagnetics Basis for Strong Force Created by Proton-Neutron-Proton- chains/rings

And Its Derived Attributes for Elements, Isotopes, and Isoforms (different N-ring-plus configurations) of Nucleus.

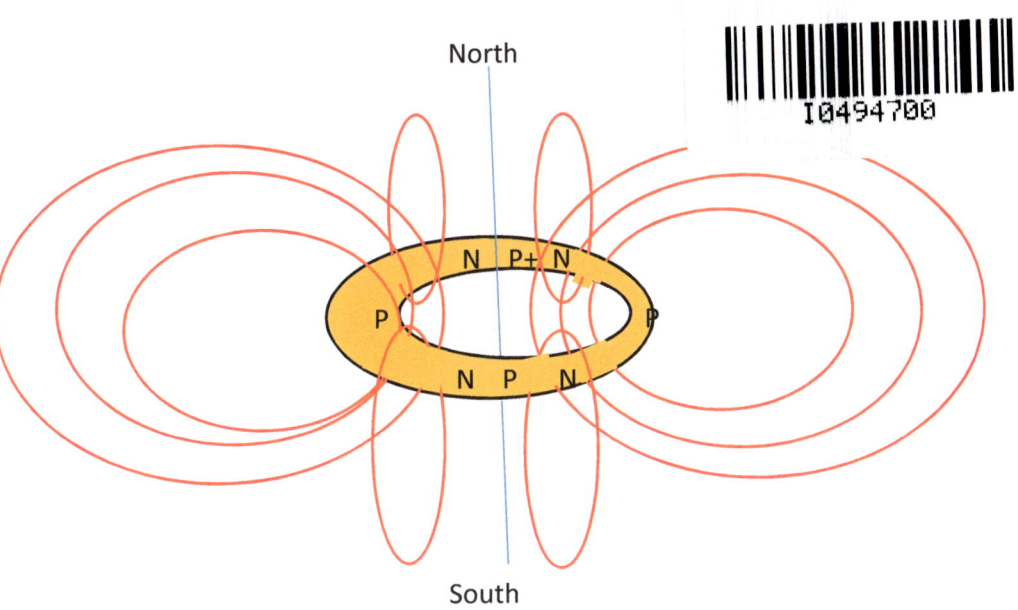

By Arno Vigen

Simple Words to Understand . . . Atoms and Chemistry

Why does a Nucleus Stay Together If Protons Repel?

- A Nucleus is Just . . . a Nucleomagnetics Ring

Why Don't Electrons Fall into the Opposite-Charged Nucleus?

- Electrons are Just . . . Frightened by Nucleus nucleomagnetics

Electron Shell Chemistry Is Just . . . Scrunched Cube Geometry

- Why are electron shells in sets of 2, then 8, then 8 and such? Can we improve Pauli-aufbau?

Scrunched Cube Periodic Chart of Elements

- What are the properties and groupings of elements using the Arno Vigen Scrunched Cube model

Scrunched Cube Chemical Bonding

- Why does Nitrogen and Oxygen have Different Bonding Angles? What drives bonding?

What Makes a Molecule Solid, Liquid, or Gas?

- And Why is the Gas of Every Element the Same Volume (a mole)?

Simple Words to Understand . . . Gravity, Electromagnetism, and Other Forces

Gravity is Just . . . That Electrons are a Little Closer

- Explaining Gravity from the basics of Electromagnetism and Explaining Why Observed Mass Changes

Does Time and Space Really Warp?

- Replacing Electron-Shell Radius for Time-Space Factors in formulas such as the General Theory of Relativity

How are Electricity and Magnetism Linked?

- Exploring the Fundamental Linkage of Charge and Magnetism

Fixing Einstein's $E = mc^2$

- Given mass $m = \dfrac{M(z,n)}{\frac{8}{3}\pi(R_{ES})^3}$ defined by AVSC nucleomagnetics, then what is mc^2

Beyond the Quantum Era

- Restoring Newton, and Moving Beyond Quantum Mechanics

Simple Words to Understand . . . Personality

Visual Astrology: Fun, Support, Security, and Growth

- Astrology 'signs' archetypes are based upon powerful traits to understand people

Visual Astrology Relationships

- What happens when 'sign' personalities interact

Visual Astrology and Jung

- Astrology 'signs' archetypes actually predict all the Jungian 4 archetypes

Dominant Personality Traits

- Dominant Personality Traits Follow from Four Dimensions, Six Steps and so 24 Subcategories

Simple Words to Understand . . . Communications

Decision Matrix® Writing

- Persuasion is based making arguments at the correct strength in a certain order.

GATESOUP® Writing

- **G**oal, **A**udience, **T**heme, Enough **E**lements, **S**upport and the rest

Kedarf® Grammar and Composition Explained

- Defining the Parts of Speech, Paragraph Structure and More in Usable Terms

Table of Contents

Challenge: What Keeps a Nucleus Together if Protons Repel Each Other?

Charge is the Most Powerful Force in the Universe

Big hugs. Let's get started exploring atoms.

Charge-force lights our cities so bright that anyone can see it, as electricity, at work from deep outer space.

Protons, with their positive electric charges, are powerful. Yet, that is a huge challenge because positive charge (+) protons both a) attract opposite charges (-) in electrons; but more importantly b) protons (+) repel like-charges; that is, other protons. *And, a nucleus consists of a bunch of protons each with positive (+) charges, and like-charges repel each other.*

Protons in the Nucleus Repel Each Strongly

Electrostatic Charge-force follows the rule:

- Like-Charges Repel

  o Positive (+) to positive (+) repels
  o Negative (-) to negative (-) repels

- Opposite-Charges Attract

  o Positive (+) to negative (-) attracts
  o Negative (-) to positive (+) attracts

Holding a nucleus together is very challenging. Under normal circumstances, the charge force far exceeds the magnetics force.

1) Its strength is substantially greater - between ~100 times stronger at the one meter apart.

Factor for 001-H <> 001-H atoms@1m	Force Calculation
Electrical Charge Force	10^{-28} m^1 / (s^2) *
Magnetic Force	10^{-30} m^1 / (s^2) *

2) Rate of Decrease of strength

Force	Rate Decrease Method
Electrical Charge Force	1/distance-squared
Magnetic Force	1/distance-cubed

So, for both attributes Charge-Force applies more:

	Force Calculation	Force Decreases
Electrical Charge Force	Better	Better (Slower)
Magnetic Force	2nd	2nd

Common sense tells the same. You don't see magnetics from space.
Charge is more powerful. You can see it.

Solving how the strongest force in the universe does not break up a
nucleus is the adventure. The basic question is: **Why does a nucleus
hold together if charge-force is so strong?**

Describing the Forces

The two strongest forces, electrostatic charge-force and nucleomagnetics-force, work in balance within a nucleus, but before I describe that relationship, I will review how electrostatic charge-force and nucleomagnetics-force are the same and how the two differ.

Charge-Force is Spherical

Charge force goes in every direction. At any distance, in any direction, the charge is the same. This applies in 3D, the X, Y, or Z direction.

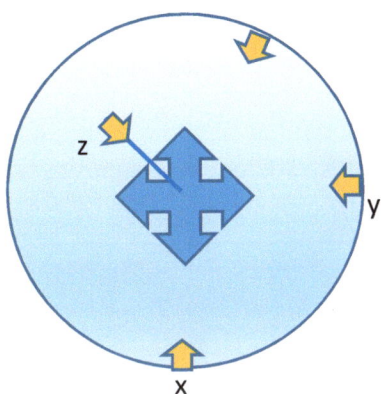

At the same distance, the field strength (yellow arrows) is the same.

Nucleomagnetics Force is North-South Oriented

Nucleomagnetics-force is north-south oriented. It is strong at 90 degrees and very weak at the poles.

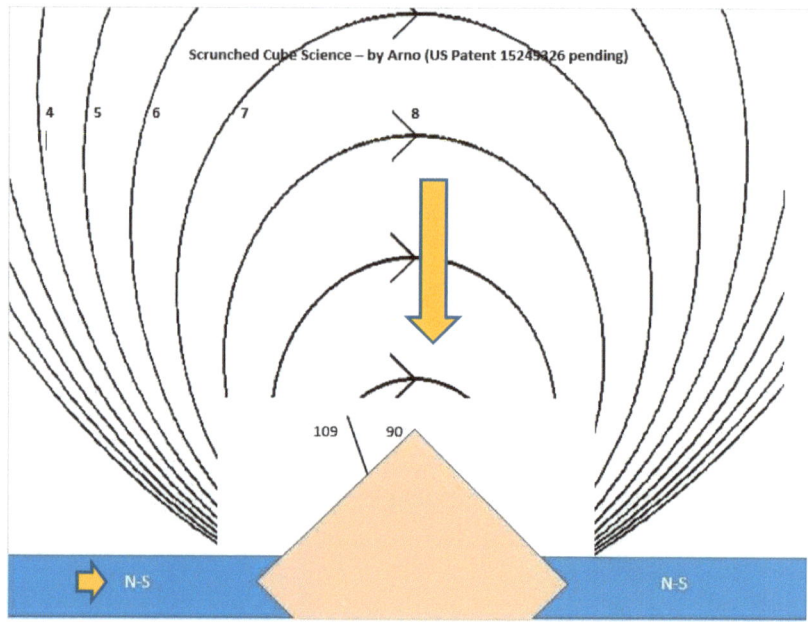

Scrunched Cube Science – by Arno (US Patent 15249926 pending)

Magnetic-force is stronger at the 90 degrees, and weaker at the nucleomagnetics poles. The nucleomagnetics field is like a 'bagel', bulging at 90 degrees to the magnet itself.

Charge-Force decreases by 1/distance-squared, Nucleomagnetics Field Strength decreases 1/distance-cubed.

The strength of a charge force comes from the proton (+) or the electron (-). Once that charge exists, its strength decreases based upon the volume.

If you have a charge at a small area 1x1x1 as a starting point. For our calculations that is:

$$Strength = \frac{2(\sqrt{k})(Q)}{(1)^3}$$

$$Force = \frac{(\sqrt{k})(Q)}{(1)^2}$$

For an example twice, as part away, then:

$$Strength = \frac{2(\sqrt{k})(Q)}{(2)^3} = \frac{2(\sqrt{k})(Q)}{8}$$

$$Force = \frac{(\sqrt{k})(Q)}{(2)^2} = \frac{(\sqrt{k})(Q)}{4}$$

The makes sense. If you have a strength of 8 at a point, then distribute that energy to a field that is 2x2x2, then the volume of 8 which makes each volume, now spread 1 in strength, with 8 boxes.

At the initial box, the strength is 8.

 1 box x 8 strength = 8 total field strength

When distributed over 8 of the same volumes, then the strength is 1 for each box (1/8ᵗʰ of the initial strength). That way the same total energy is conserved. Eight (8) boxes of 1 is the same as one box worth eight (8).

 8 boxes x 1 strength = 8 total field strength

Now the real answer is a sphere, with $\frac{4}{3}\pi r^3$, but the simplified cube as a picture is hopefully enough to show the concept that a charge distributes evenly across the volume as it (the field strength) spreads to a distance.

So, for a position at radius two, that field has the same strength, but it is distributed over the entire volume. Therefore, at 2x (twice) the distance, the field is 8x stronger ($\frac{1}{d^3}$).

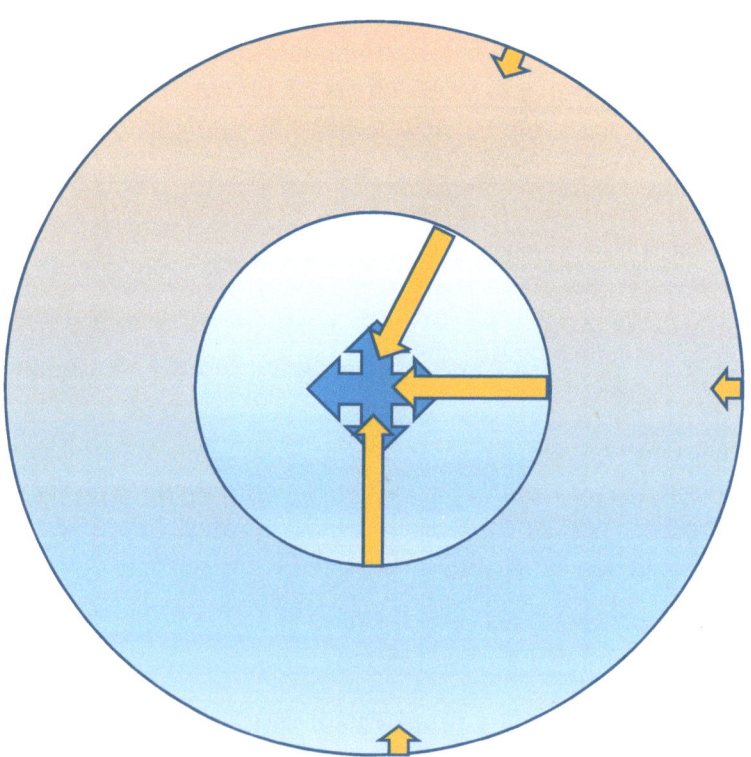

Yet, if you are looking at this from one direction, then that one strength factor does not decrease, you get all the strengths in that direction. Therefore, the force decreases by 1/distance-squared ($\frac{1}{d^2}$)

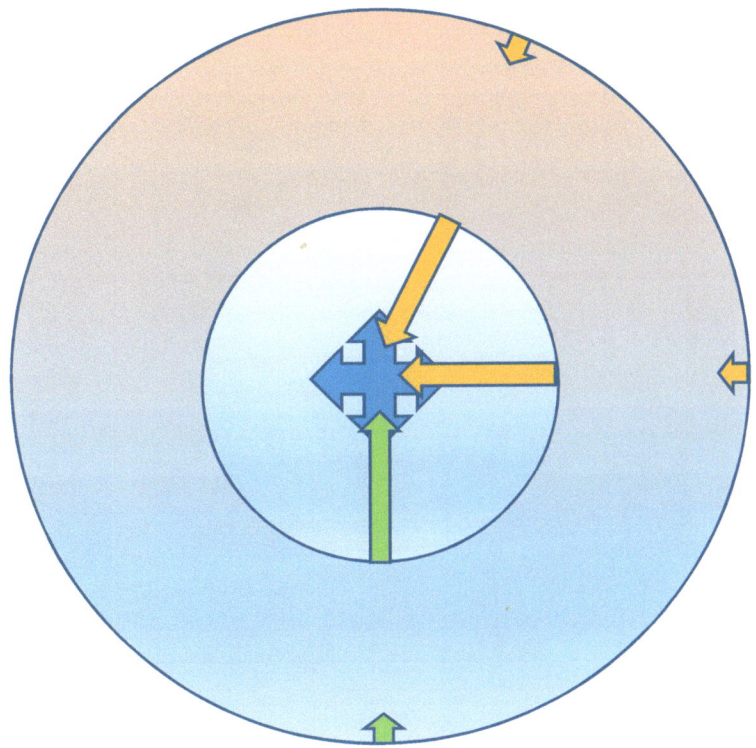

Think of it this way. In the direction of the distant object, all those strengths (green arrows) are still applying fully. So, in one dimension, the field strength does not decrease. That leaves only two (2) dimensions (a factor with base $\frac{1}{d^2}$) that decrease with distance **_relative to_** a distance object in one direction.

A nucleomagnetics field is already has a decrease by the oriented direction (purple arrows).

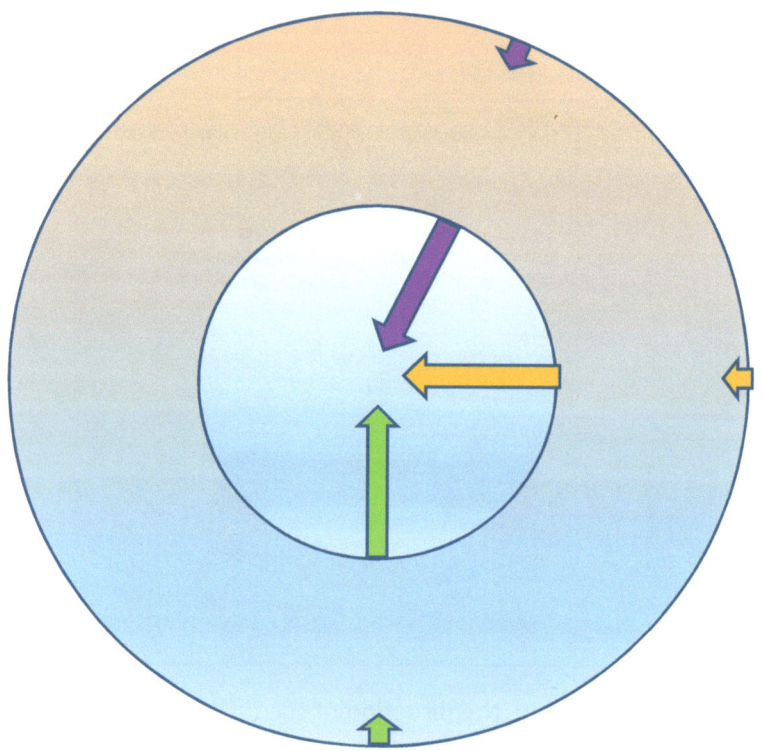

Therefore, from that direction is a) already low, b) decreases by $\frac{1}{d^2}$. Yet, in the other directions, the nucleomagnetics force gets pulled down by orientation dimensions, so $\frac{1}{d^2}$ becomes $\frac{1}{d^3}$ based upon the angle after the end of the magnet. Or you can think of that as

$$\frac{1}{(d^2)(d*\arcsin\ )}$$

16

The nucleomagnetics field starts decreasing that edge of the particle, and for out purposes, that is a tiny sphere in the middle of the field.

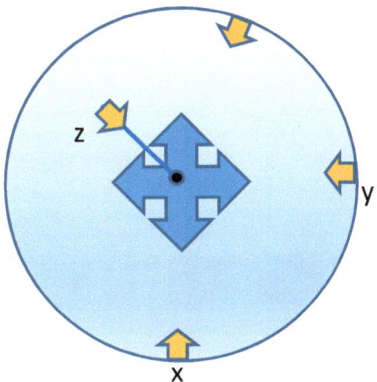

Charge-force decreases in every direction, and thereby, all of the field vectors point to the center of the origin particle.

The center of a charge-particle is also the center of the charge-force. Therefore, even if the particle has dimensions (X, Y, Z), that charge-force centers in the same place as center of the particle.

Yet, that is not the case for nucleomagnetics forces.

The nucleomagnetics field remains the same along the width of the magnet. It does not decrease its strength until it gets to the end of the magnet.

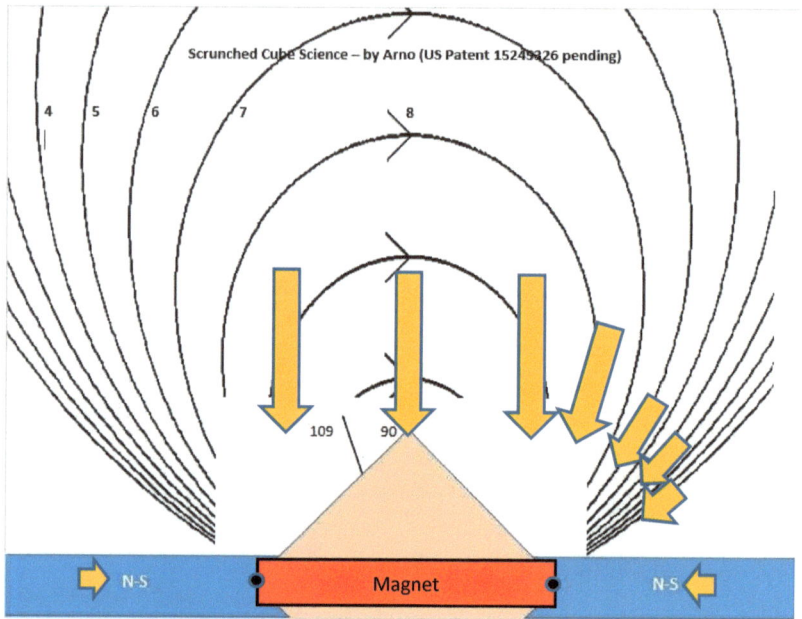

Further, the field points toward the nearest pole, not the actual center of the magnet. The field vectors point to a place different that with electrostatic charge-force.

The nucleomagnetics field really points to each of the ends of the magnet when past that end. In the middle, they point to perpendicular – toward the nearest point on the nucleomagnetics bar.

In this tiny way, electrostatic charge-force and nucleomagnetics-force are different.

Summary:

Therefore, in terms of initial strength, Charge-force is stronger:

Factor for 001-H <> 001-H atoms@1m	Force Calculation
Electrical Charge Force	10^{-28} m^1 / (s^2) [ii]
Magnetic Force	10^{-30} m^1 / (s^2)

Therefore, in terms of rate of decrease, Charge-force is again stronger:

	Strength Decrease Factor
Electrical Charge Force	$\dfrac{1}{d^2}$
Magnetic Force factor	$\dfrac{2^{\frac{1}{2}}}{d^3}$ at 90 degrees $\dfrac{1}{d^3}$ at nucleomagnetics poles, And $\dfrac{\sim\left(1+3COS(\theta)^2\right)^{\frac{1}{2}}}{(d^3)}$ between

So, for both attributes Charge-Force applies more:

	Force Calculation	Force Decreases
Electrical Charge Force	Better	Better (Slower)
Magnetic Force	2nd	2nd

With that background, we are ready for answer the basic question:

Why does a nucleus hold together if charge-force is so strong, both a) in strength and b) in its lower rate of decrease?

> This book will not explore the underlying relationship of charge to magnetism. That will need an entire book of its own. That question is the subject of countless hours of research and speculation. That relationship is fundamental and itself very interesting.

Magnetics of Nucleus Particles in Chains Causes Enough Nucleomagnetics Attraction to Overwhelm Proton-Proton Charge Repulsion

In the rest of this proof, we will move between applying a) the rules that apply to, and b) the outcomes that result from, these two most-powerful forces as we build the basic structure of a nucleus, and then the nuclei of different sizes, shapes, and elements within the periodic chart and their isotopes.

First, the problem is strength. Electrostatic Charge is stronger at observed distances.

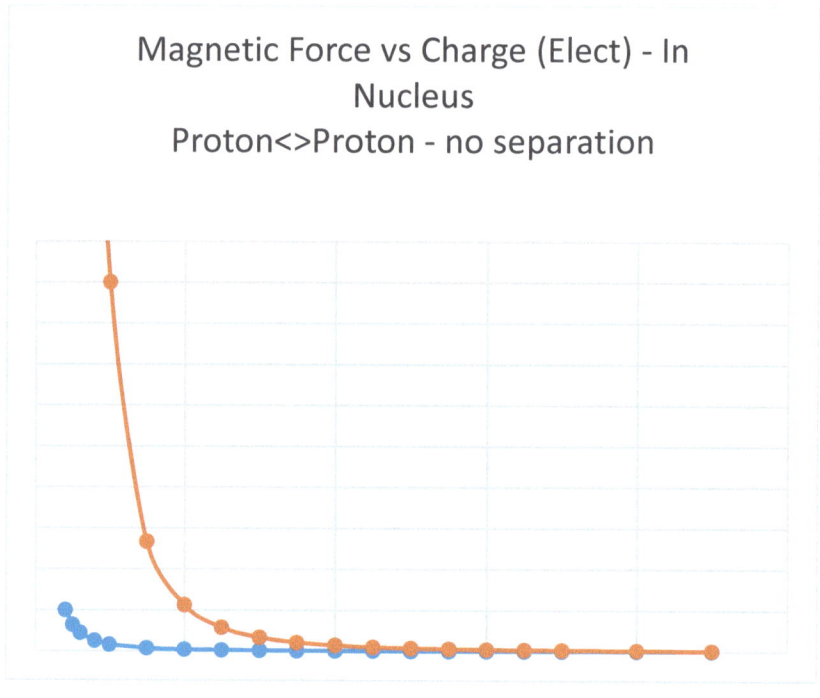

Magnetic Force vs Charge (Elect) - In Nucleus
Proton<>Proton - no separation

The repulsive force of same-charge between two protons diminishes based upon the square of the distance ($\frac{(\sqrt{k})(Q)}{d^2}$). The attractive force of magnetism diminishes based upon the cube of the distance ($\frac{M}{d^3}$) depending upon orientation. [iii]

If you throw protons fast enough to penetrate a nucleus, and thereby touch proton-to-proton, you create a nuclear reaction. That repulsive proton-proton force created is explosive. The 1/distance-squared at small distances nearing ($\frac{1}{0}$) become exponentially devastating – infinite force.

If there was nothing else, a nucleus would not stay together. The protons would push each other away. Then why does a nucleus stay together?

The above looks at those two curves of distance versus force shows that the Electrical charge (repulsion in a nucleus) in red/orange always exceeds the nucleomagnetics force in blue if the two forces are at the same place.

But, there is an amazing additional fact:

A Magnet Stays Strong if in a Chain

We already discussed that nucleomagnetics do not start decreasing until the end of the magnet. That the nucleomagnetics-force points to the poles, not the center of the magnet.

This applies when particles, by their nucleomagnetics, bind N-S-N-S. The new combination is a full magnet, and again the field decrease does not start until the end of the 'combined nucleomagnetics structure' - at the ends.

The nucleomagnetics force extends if physically connected. It does not start decreasing until the end of the magnet. That is, a chain of magnets keeps the force at the same strength – while the proton charge repulsion keeps decreasing. When separated by a neutron, you chain particles in the nucleus neutron, protons, and so on, so that the nucleomagnetics field decrease starts further to the right; the blue line stays flat for that distance before it starts its decrease.

Let's see how that works. Remember that this movement of the charge strength to additional distance only occurs if physically connected. In this case, that connection is a proton-neutron-proton nucleomagnetics chain. I move the nucleomagnetics-force (blue) line in the chart.

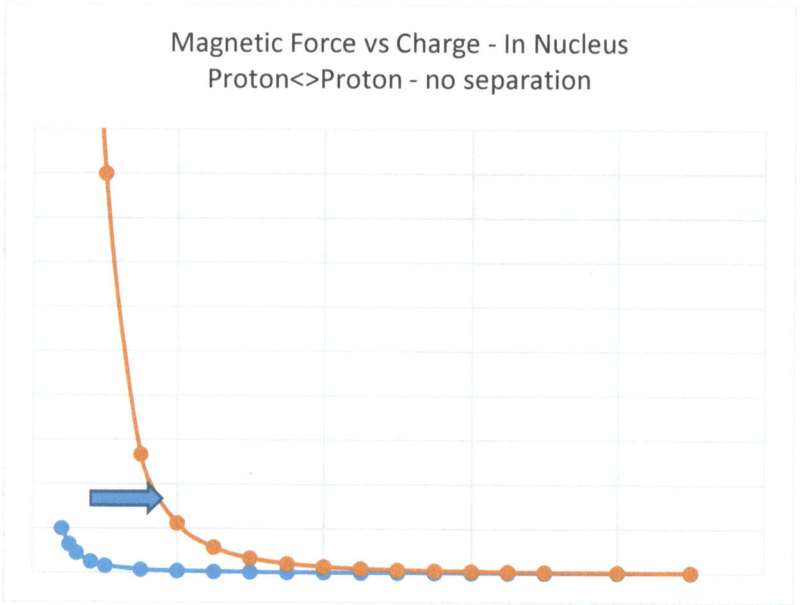

This shows that the nucleomagnetics force and charge force sometimes trade places. When the charge is moved away, but the magnet-in-chain/still intact remains.

At very short distance, both charge and magnet become huge. However, there is a limit of the physical thing. A magnet builds in the object so someplace at the left, the curve ends. It cannot become infinite because the actual magnet structure is not zero in length. These curves are not applicable at less than the actual size of the particles involved.

There is a place in the middle where the nucleomagnetics force is stronger than the charge force.

That new curve with both forces looks like below and for a range the nucleomagnetics (blue) force stays above the charge (orange) force:

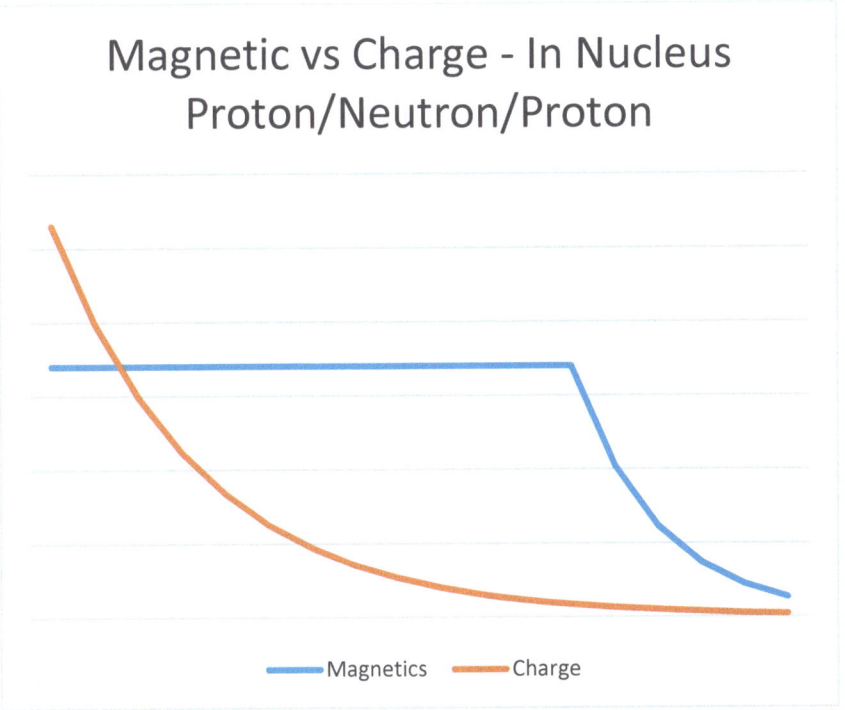

The nucleomagnetics force (blue line) I show as constant along a continue chain of nucleomagnetics entities which ends at the length of the proton or neutron (about .84 to .87 x 10^{-11}). After that, the curve decreases by the $F = \dfrac{M}{(d^2)(d*\text{arcsin })}$ such that eventually the charge repulsion is greater at some distance from the nucleus. The (blue line)

Charge force is distance-squared and larger, but only at location closer than the neutron separator.

Now, the above graph shows the two forces, but the charge is repelling and the nucleomagnetics is attracting. So, really you need to look at them as the net. That is the size of that different (the space between the two functions which is sometime positive (net repulsion) and sometime negative (net attractions).

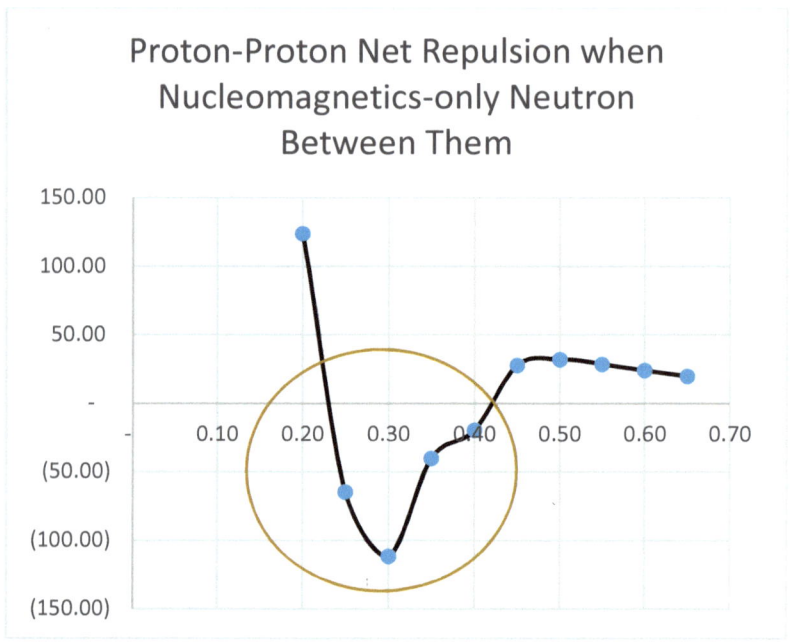

Proton-Proton Net Repulsion when Nucleomagnetics-only Neutron Between Them

Compare the circled areas to this graph of a gluon to the current Wikipedia (the hump to positive is the electrostatic charge taking over at a distance:

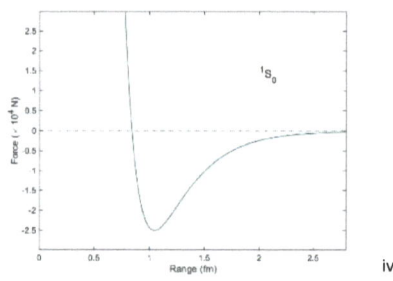

iv

24

At too close, the structure repels (explosive nuclear reaction force). At far away, the structure also repels (magnetic becomes weaker, faster by 1/distance-cubed). However, at a magic distance in the middle, the structure actually attracts (the area in the middle below zero).

If you look at current science books or Wikipedia about 'strong interaction' or 'strong force', they describe this as a particle called a gluon, which has the exact force shape as the negative (attractive). In the AVSC (Arno Vigen Scrunched Cube) Atomic Model, it is portion of the net-force of electrostatic and nucleomagnetics force shown above, and no gluon particle exists.

For the Advanced explorer:

For this balancing at long distances, there is a more detailed analysis in my Gravity is Just . . . book. At long distance, the nucleomagnetics force is 1/distance cubed (M/d^3) depending on orientation so it becomes smaller than 1/distance-squared (C/d^2). At long distances, only charge is important. The nucleomagnetics-force decreases faster, so nucleomagnetics force only applies close to the atom.

Further, the nature of 1/distance-cubed actually has interesting change to stronger. In fact, it is stronger inside the Bohr radius, but only up to the point of the distance of a subatomic particle. While I start with the common incomplete idea that electrostatic charge is strong, that fact is only at distances beyond the atom. At subatomic distance, nucleomagnetics is the most powerful, and drives the AVSC normalizing calculation make a determinate methodology parallel to and in any calculation requiring geometry, superseding Schrodinger's Equation.

Magnetic Neutrons are the Required Separators Extending nucleomagnetics While Charge Decreases

What is needed then is a magic particle. A particle that:

- Has no charge

- Is nucleomagnetics strength and orientation

That particle exists; it is a neutron.

With a neutron, atoms get created that have the basic need for building a nucleus structure that is stable. It is a structure proton-neutron-proton and so on . . .

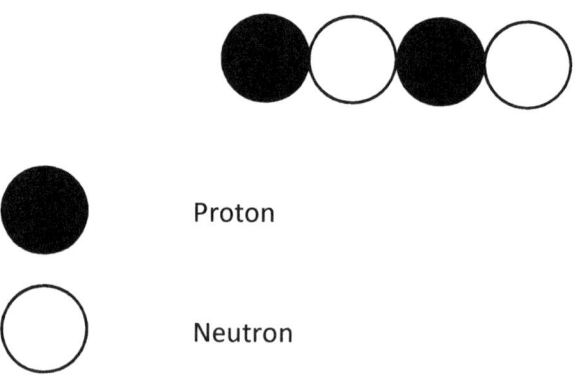

Proton

Neutron

So long as the structure builds from this logic, then a stable nucleus can exist. The distance of a neutron separates enough to lower the charge-force, AND the physical connection maintains the nucleomagnetics-force of the chain of nucleus particles.

Calculation of the Strength of Electrostatic Charge and the Strength of Nucleomagnetics Forces at Various Distances of the Radius of a Neutron

There are three basic factors as work in this nucleus binding, the 'strong nuclear interaction' or 'strong force' per most textbooks: the proton-proton repulsive force, the oriented nucleomagnetics field (attractive when north to south), and the actually physical dimensions of the particles.

1.

Proton-proton repulsion is determined using Coulomb's Law, which is generally understood as decreasing by 1/distance-squared ($\frac{1}{d^2}$).

$$F = k_e \frac{q_1 q_2}{d^2}$$

2.

The oriented nucleomagnetics field (attractive when north to south) is the most difficult, but it is generally understood as decreasing by 1/distance-cubed ($\frac{1}{d^3}$).

$$F = M_A \frac{(z+n)_1 f(\theta_1)(z+n)_2 f((\theta_2)}{d^3}$$
$$= \frac{M_{root-} (z+n)_1 f(\theta_1) M_{root-} (z+n)_2 f(\theta_2)}{d^3}$$

3.

And, the actually physical dimensions of the particles. Which are actually limits to the distance (d^2). The nucleus particles (protons or neutrons) are generally understood to have dimensions of:

Nucleus	0.8751(61) fm[1] (0.8751×10^{-15} m)[V]
Charge radius	

Calculation of Magnetism by Comparison to Charge at Bohr Radius

As noted in Book 1 of Simple Words to Understand . . . Gravity, Electromagnetism and Other Forces: <u>Gravity is Just That . . . Electrons are a Little Closer</u>, charge-force and nucleomagnetics-force are intricately linked.

At approximately the Bohr radius, the balancing of those two creates the placement of the electron shell. At that distance (d^2), the attraction electron-proton charge-force equals the electron-magnetic field of nucleus repulsion.

The Bohr radius is:

Bohr radius · a$_0$
$5.2917721092171717 \times 10^{-11}$m r = n^2 a$_0$/Z [vi]

So,

$$F = k_e \frac{q_1 q_2}{d^2}$$

Factor	Charge-Force Estimation
Factor	**Charge-Force Estimation**
Charge-Force Constant (Coulomb)	$k_e = 10^{10}$ [8.99×10^9 m^3 kg^2/ (s^2)]
Charge 001-H Hydrogen atom = 1 proton	$q_1 = 10^{-19}$ [1.67×10^{-19}]
Charge 001-H Hydrogen atom = 1 proton	$q_2 = 10^{-19}$ [1.67×10^{-19}]
Distance	d=5.29 x 10^{-11} m or 10^{-10} m (d^2=5.29x10^{-11})^{2}m =10^{-21} m
Exponent shortcut	+k+Q+Q-d-d
Short-cut calculation	10-19-19-(-21)= 10-19-19+21= 10-38+21=-7
Charge-Force Repulsion	10^{-7} m^1 / (s^2) (Newtons)

Yet that has to be equal to the nucleomagnetics repulsion at that same distance.

$$F = M_A \frac{(z+n)_1 f(\theta_1)(z+n)_2 f(\theta_2)}{d^3}$$

Factors	Magnetic-Force Estimation
Charge-Force Constant (Coulomb)	$M_A = 10^{-38}$ $m^3 \, kg^2/(s^2)]$
Charge 001-H Hydrogen atom = one (1) Proton	n=1 or 10^0
Charge 001-H Hydrogen atom = one (1) Proton	n=1 or 10^0
Angle factor-1st Hydrogen	1
Angle factor-2nd Hydrogen	1
Distance	d=5.29 x 10^{-11} m or 10^{-10} m
	(d^3=5.29x10^{-11})^{3}m =10^{-31} m
Exponent shortcut	+k+Q+Q+0+0-d-d-d
Short-cut calculation	-38+0+0-(-31)= -38+0+0+31= -38+0+31=-7 N (Newtons)
Charge-Force Repulsion	10^{-7} m^1 / (s^2) **(Newtons)**

Measuring Charge-Force versus nucleomagnetics-Force at Distance in Nucleus

For the postulate to work, at the levels of the nucleus particles, the nucleomagnetics-force must exceed the charge-force.

Of course, given the two are equal at the Bohr radius, and the magnetism will inverse at cube versus the square of charge-force, one would expect that magnetism is must stronger at the distance that of nucleus particles.

That distance would be 2 x radius since it must transfers the one particles and attach to the 2nd.

Let's test this for proton-proton charge-force repulsion at that separation and then for proton-neutron nucleomagnetics-force attraction at that separation.

The charge force at a nucleus distance would be

$$F = k_e \frac{q_1 q_2}{d^2}$$

Factor	Charge-Force Estimation
Charge-Force Constant (Coulomb)	$k_e = 10^{10}$ [8.99x10⁹ m³ kg²/ (s²)]
Charge 001-H Hydrogen atom = 1 proton	$q_1 = 10^{-19}$ [1.67×10⁻¹⁹]
Charge 001-H Hydrogen atom = 1 proton	$q_2 = 10^{-19}$ [1.67×10⁻¹⁹]
Distance	d=0.86 x 10⁻¹⁵ m or 10⁻¹⁵ m
Exponent shortcut	*+k+Q+Q-d-d*
Short-cut calculation	*10-19-19-(-15)-(-15)=* *10-19-19+15+15=* *10-38+30=+2 N (Newtons)*
Charge-Force Repulsion	**10⁺² m¹ / (s²) (Newtons)**

Yet that has to be less than the nucleomagnetics repulsion at that same distance – which it is.

$$F = M_A \frac{n_1 n_2}{d^3}$$

Factors	Magnetic-Force Estimation
Charge-Force Constant (Coulomb)	$M_A = 10^{-38}$ m³ kg²/ (s²)]
Charge 001-H Hydrogen atom	n=1 or 10⁰
Charge 001-H Hydrogen atom	n=1 or 10⁰
Distance	d=0.9 x 10⁻¹⁵ m or 10⁻¹⁵ m
Exponent shortcut	*+k+Q+Q-d-d-d*
Short-cut calculation	*-38+0+0-(-15)-(-15)-(-15)=* *-38+0+0+15+15+15=* *-38+45=+7 N (Newtons)*
Magnetic-Force Attraction	**10⁺⁷ m¹ / (s²) (Newtons)**

So, nucleomagnetics-force is larger compared to charge-force at the same nucleus particle distance. That is, if the particles are separated by the distance of a particle.

However, the real comparison is that charge-force, if proton-near-proton is much less. If no separating particle, the protons get much closer as shown below. The proton-proton repulsion is huge, even versus the nucleomagnetics attraction.

$$F = k_e \frac{q_1 q_2}{d^2}$$

Factor	Charge-Force Estimation
Charge-Force Constant (Coulomb)	$k_e = 10^{10}$ [8.99x10^9 m^3 kg^2/ (s^2)]
Charge Proton	$q_1 = 10^{-15}$ [0.9×10^{-15}]
Charge Proton	$q_2 = 10^{-15}$ [0.9×10^{-15}]
Distance	d=3.0 x 10^{-16} m or 10^{-16} m
Exponent shortcut	*+k+Q+Q-d-d*
Short-cut calculation	*10-16-16-(-10)-(-10)=* *10-16-16+15+15=* *10-32+30=8 N (Newtons)*
Charge-Force Repulsion	**10^{+8} m^1 / (s^2) (Newtons)**

If there is not that neutron separation, then the charge force can get too close, and become 10^{+8} (~10,000,000) m^1 / (s^2) vs the nucleomagnetics attraction of 10^{+7} (~1,000,000) m^1 / (s^2) of proton-proton.

If there is the neutron separation, then the charge force is overcome because the 10^{+2} (~100) m^1 / (s^2) charge-force when separated by a neutron is then less than the nucleomagnetics attraction of 10^{+7} m^1 / (s^2) of proton-proton.

As such, the forces at play do not bond proton-proton, but do bond proton-neutron-proton.

Ultimately, the particle physics of what goes on at those levels, down to quarks, is beyond the scope of this postulate. That would take years more to study and understand.

Yet, it is enough from the above calculations using just electrical-charge-force and magnetism-force that magnetism is stronger at the nucleus distance level; strong enough to hold a nucleus together at the distance of proton-neutron-proton separation, and yet charge-force could be much bigger and creates observed nuclear explosion without that separation if the proton come near to touching.

Configurations of Nucleus Chains/Rings

Basic Chain

The nucleomagnetics cause the top ring to flow P > N > P and so on. There is a nucleomagnetics pole (blue) and nucleomagnetics fields (red) around it.

A 001-H Hydrogen is easy because it is just one proton, so there are no other proton (+) particles to cause repelling. However, 001-H Hydrogen also has two isotypes which include neutrons.

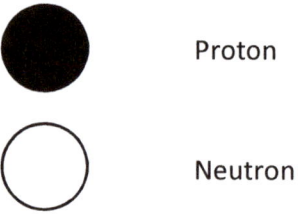

Proton

Neutron

001-H Hydrogen Deuterium (2 Atomic Weight Isotope)

001-H Hydrogen Trillium (3 Atomic Weight Isotope)

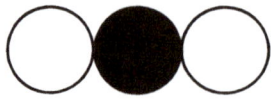

002-He Helium (3 Atomic Weight Isotope)

002-He Helium (4 Atomic Weight Isotope)

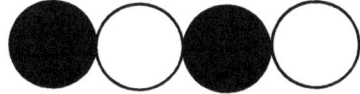

These structures are stable. It meets the primary rule which is:

At least one neutrons must separate every pair of protons (+).

A nucleus could continue with this logic and create these long chains.

This structure has a long chain, and that makes the structure have a nucleomagnetics orientation.

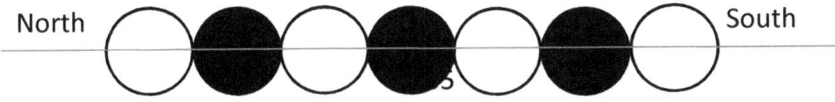

North   South

Remember that balancing act of 1) charge and 2) nucleomagnetics logic throughout this proof.

This would be great except that a chain is not rigid. A chain can bend around, and maybe those two ends meet again – and BANG a proton-proton nuclear-reaction repulsion.

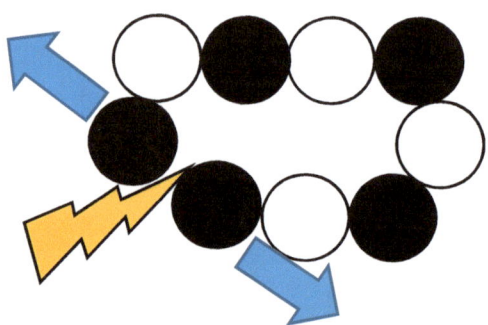

If a nucleus has a long chain and the two ends create proton-proton nuclear-reaction, which makes the chain not the most stable.

Magnetism-force follows the rule about attraction and repulsion similar to the charge-force except that the attribute is the direction of the magnet – north versus south:

- Like nucleomagnetics Poles Repel

  o Positive (+) to positive (+) repels
  o Negative (-) to negative (-) repels

- Opposite nucleomagnetics Poles Attract

  o Positive (+) to negative (-) attracts
  o Negative (-) to positive (+) attracts

So, if the two ends are a proton and a neutron, then 1) there is no proton (+)-proton (+) charge repulsion; and 2) the neutron adds nucleomagnetics attractions; which is 3) by the chain the attractive north-south orientation. Therefore, a chain if ending one end with a proton and the other end with an electron has a high tendency to flow around and become a ring.

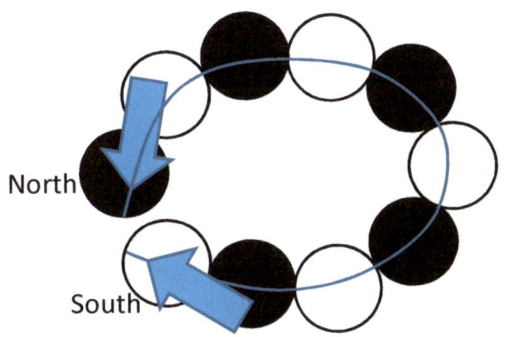

So, there is a very stable alternative for Helium.

002-He Helium (4 Atomic Weight Isotope)

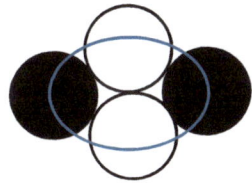

By the way, notice that the neutrons want to get closer, versus the protons, to make the proton separation as great as possible while still maintaining that looped magnetic chain, a ring structure.

A chain must have one end with a north pole, and the other with a south pole. That means that, unless both end with protons, the two ends want to come together. Even if both ends are neutrons, the chain will still join.

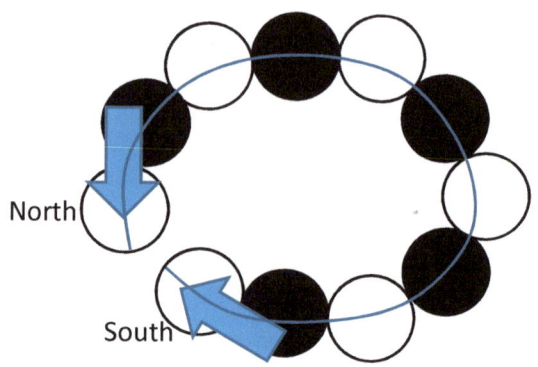

So, most often, the nucleus chain will still join into a ring. That structure is stable, and no endpoints can 'flap' and create the nuclear-explosion repulsion.

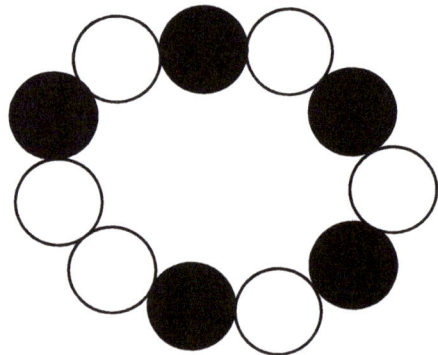

Corollary: More than one neutron can separate every pair of protons (+).

Finally, there can be combinations of rings and chains. The building block do not work neatly or perfectly.

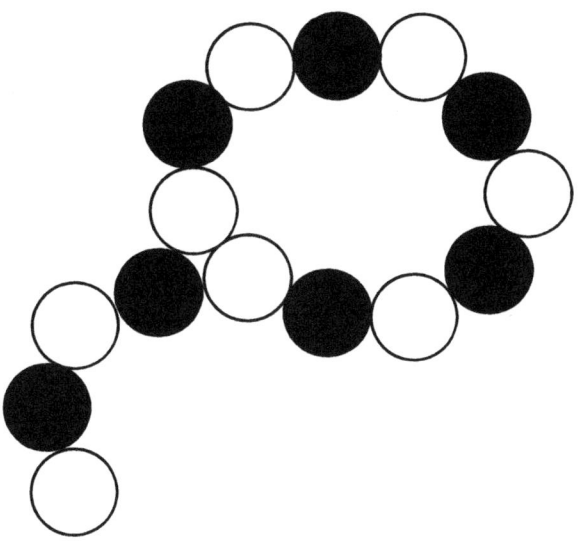

These combinations can also go in three dimensions (3D) which is more difficult to show on a 2D page.

These complex combinations become more important as you get to elements with lots of particles, especially those that are radioactive. However, the next chapters will focus on the stable lower end of the periodic chart with less particles. That will build the concepts, their attributes, and their impact on chemical structures, bonding and chemical reactions.

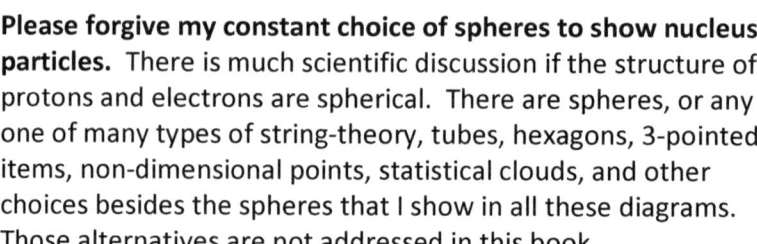

Please forgive my constant choice of spheres to show nucleus particles. There is much scientific discussion if the structure of protons and electrons are spherical. There are spheres, or any one of many types of string-theory, tubes, hexagons, 3-pointed items, non-dimensional points, statistical clouds, and other choices besides the spheres that I show in all these diagrams. Those alternatives are not addressed in this book.

I am using spherical particles simply because the pictures with spheres work to make ring angles easier. Please indulge me with that visual concept. I am not claiming that all nucleus particle are spheres; I just like them for the picture of the ring-concepts.

Much research describes small-number-of-particles atoms have a nucleus size at $0.8 * 10^{-15}$ m, yet larger-number-of-particle atoms have a nucleus size at $1.5 * 10^{-15}$ m suggests that the distance between nucleus particles, and thereby the radius of individual particles, or subparticle clouds, actually compresses. Given that $1.5/0.8 \sim= 2$ or twice as big, you cannot have the 200 particles of a large atom fit into a volume that is only $2^3 = 8$ times bigger if the particle remains the same spacing or size. That is, the particles get into a structure which makes them fit together tighter.

My current thinking is that the volume of larger atoms for 200 particles shows the basic structure. So, at $(1.5 * 10^{-15})^3$ m^3 / 200 particles =

$$V = \frac{(1.5*10^{-1})^3}{200} m^3 = \frac{3.4*10^{-45}}{200} m^3 = 1.7 * 10^{-47} m^3.$$

Then taking the cube-root of that is the core radius of $0.3 * 10^{-15} m^3$. This corresponds with other observations of the neutrons which states it has a positive-charged core of the same $0.3 * 10^{-15}\ m$.

None of that actually resolves sphere versus cone versus string versus subparticle cloud versus something else. Therefore, I stay with sphere until something is proven.

Secondary Understanding: There can be two or more neutrons in a row.

While the perfect nucleus would be these proton and neutron in perfect order, that is not reality. Depending on the elements, you are almost as likely to have either chains or rings with places with more than one neutron together – still separating any repulsions from protons (+) touching or near-touching protons (+).

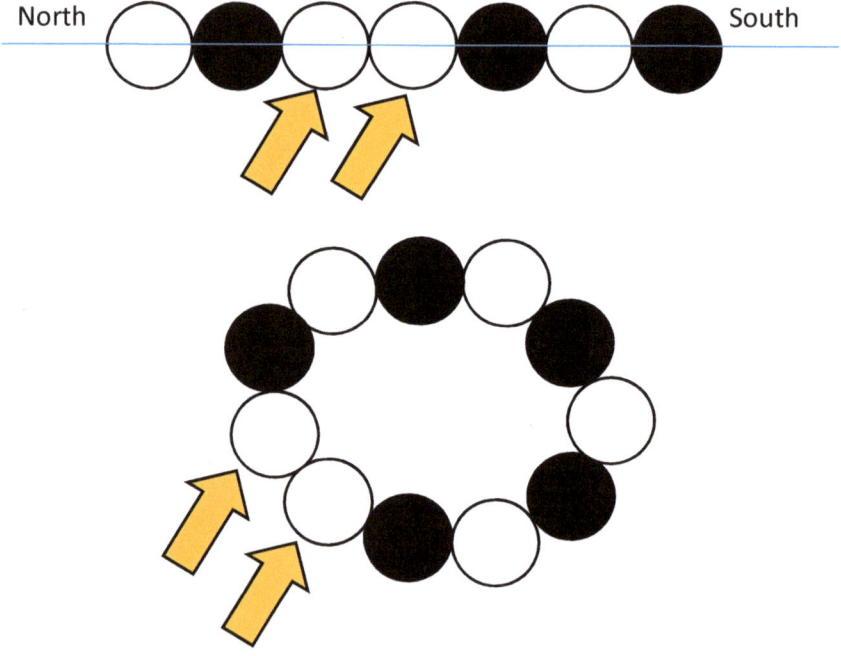

North   South

Remember this fact because it explains various forms of the same element called isotopes. That is where an element can exist with the same number of protons, but with a different number of neutrons. We will explore these isotopes more in a later chapter.

Magnetics of Entire Nucleus Structure Are Important and Different than Particle nucleomagnetics

The basic concept of a ring of nucleomagnetics particles creates a 2nd overall nucleomagnetics direction. There is a nucleomagnetics pole at 90 degrees to the ring.

The nucleomagnetics particles many orient with each south connecting to the north of the next particle. However, this creates a secondary nucleomagnetics field that has it poles through the middle of the ring. The particles have a nucleomagnetics orientation that flows like the blue arrow. However, magnets in a ring create another north-south nucleomagnetics field at perpendicular (the red line).

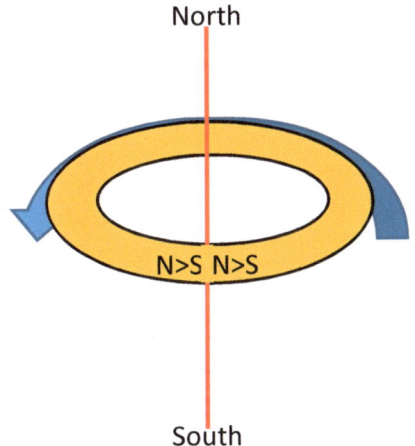

And the fat part of the filed around the ring-circle, sort of like a 'bagel' or maybe a series of these loops of nucleomagnetics field surrounding the ring. That field being strongest around the ring.

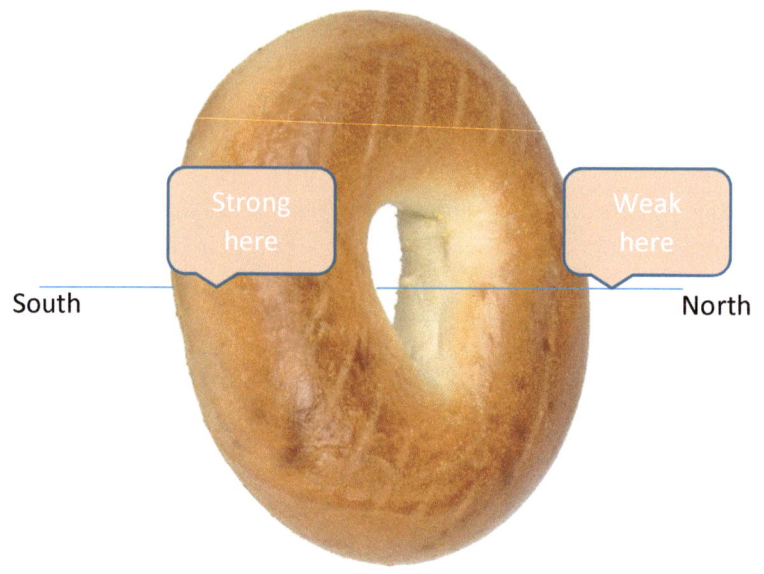

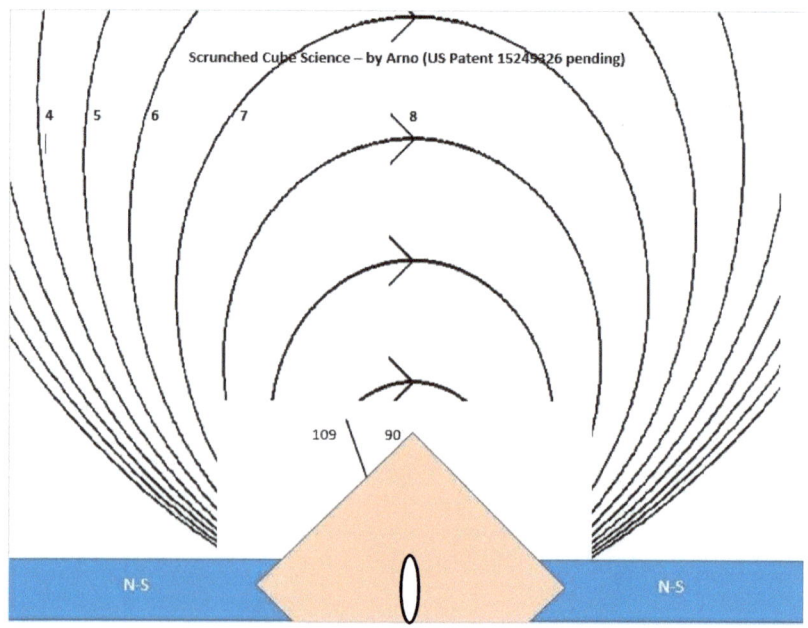

However, a strange thing happens when if the nucleus builds into two rings. The nucleomagnetics of the 'lower' ring reverses. To work, the upper must be N>S, and the lower S>N with the connection as little loops of nucleomagnetics energy.

North

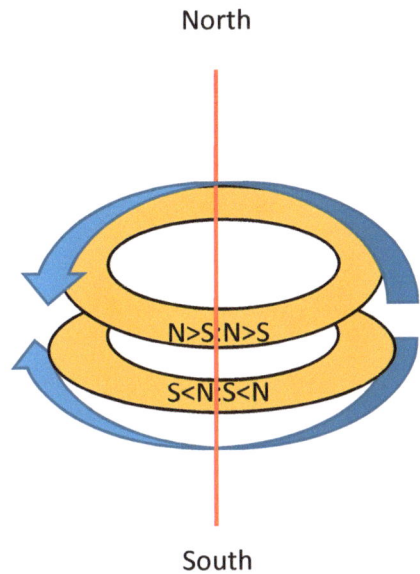

N>S:N>S

S<N:S<N

South

The nucleomagnetics field between rows becomes a series of loops.

Single-ring:

The nucleomagnetics cause the top ring to flow P > N > P and so on.
There is a nucleomagnetics pole (blue) and nucleomagnetics fields
(red) around until a ring completes.

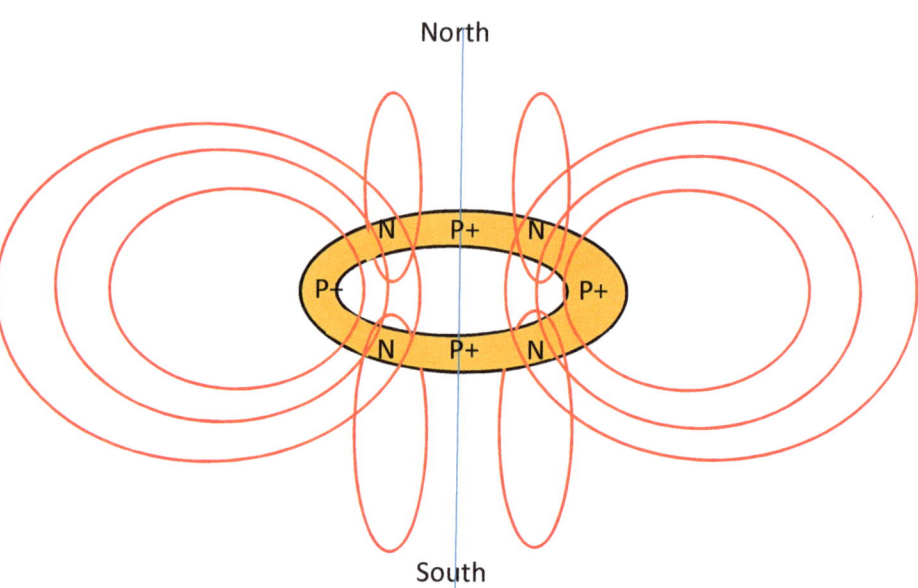

Double-ring:

The nucleomagnetics cause the top ring to flow P > N > P say in N > S

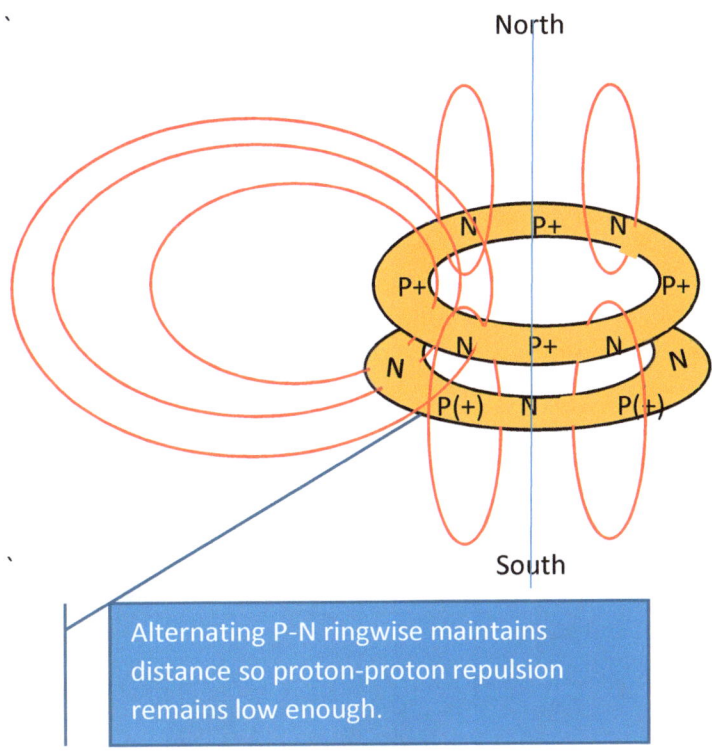

Alternating P-N ringwise maintains distance so proton-proton repulsion remains low enough.

But because of the connection from layer 1 to layer 2, the polarity changes for the 2nd layer

The nucleomagnetics cause the 2nd ring to flow N > P > N say in S > N. That is the flow of N-S is the opposite.

Building Higher Elements Needs Neutron First to Separate for Added Proton

The building of chains would seem easy. You add a neutron, then a proton fits in between. However, this of how long and wiggly this structure becomes. In fact, too much wiggly, and the chain will bend, and if a proton nears a proton, the structure explodes. Yes, a proton-to-proton touch or near-touch is the basis of nuclear explosions. Anytime that protons get too close the repulsive power becomes huge, likely tearing apart one section of the atom's nucleus from the rest.

The process has to follow a particular path of a) add neutron first, then, add proton to keep the proton-proton near-touch decay of the entire structure from occurring.

The Stages

A chain (segment of nucleus particles) is naturally proton-neutron-proton and so on.

A neutron can attach to that segment. That neutron protrudes.

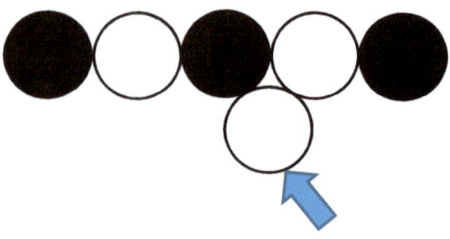

However, then the particles realign so a neutron-neutron block occurs.

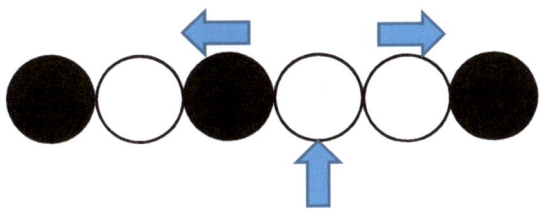

Only after that realignment, a proton can attach. That proton protrudes.

Of course, those can then re-orient back into a longer chain (and new element).

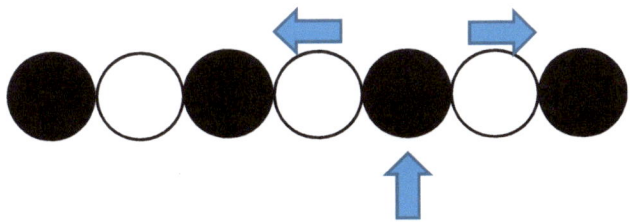

In the physical world, you will find a nucleus in each configuration. Effectively, a nucleus can be at any of the stages for any particular segment. That is why larger molecules are not perfect 2:1 ratio. It can work with a small structure, but as the structures get bigger some of these neutrons gets stranded.

	Atomic Number	Ratio
002-He Helium	4.0026	2.0013000
006-C Carbon	12	2.0000000
012-Mg Magnesium	24.3	2.0250000
092-U Uranium	238.0	2.5869565

In Multiple-Ring Structures, Extra Neutrons Keep the Protons Separated

In fact, to get to the multi-layer rings, there needs many extra neutron spacers so protons do not touch. Once an atom gets into multiple rings, the process is more complex. The 2nd layer cannot have protons in the same positions as the initial row.

That would create a nuclear explosion. However, if those align perfectly, then a 2-layer double ring works.

Side View

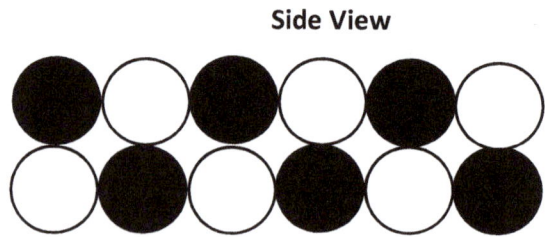

However, in nature, the layers of a nucleus ring more likely building includes some spacers. That is, a neutron sits in a spot every so often because that can happen as easily as a proton (+).

50

Side View

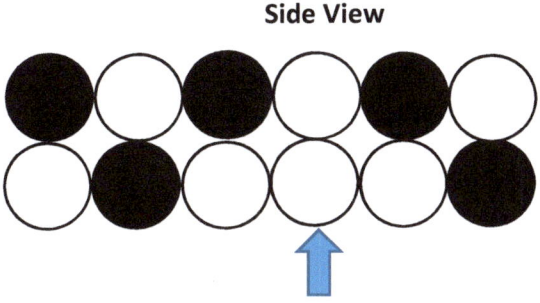

Structure Prefers Same Number of Particles on Every Layered Ring

If two layers have the different number of particles, then nucleomagnetics like to drive the structure into the slots. However, that makes every particle from one ring touch two particles in the next layer. That makes it almost impossible for a proton not to get near another proton. Of course, that, BANG, would create nuclear decay.

Side View

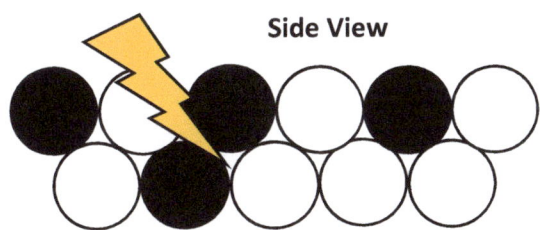

If there is an odd number of particles in each ring, then definitely an extra neutron fits into the structure to make the structure have the same number of elements on each ring-layer.

For example, 009-N Nitrogen has an isoform-isotope which has five protons (+) on one layer and four (4) protons (+) on the other layer of a 2-layer ring structure. But that does not work. The firth positions shifted alternating would become proton at each end where the rings connect. Instead, one common isotope with one extra neutron makes the structure stable. This is a P-9, N-10 isotope.

Side View

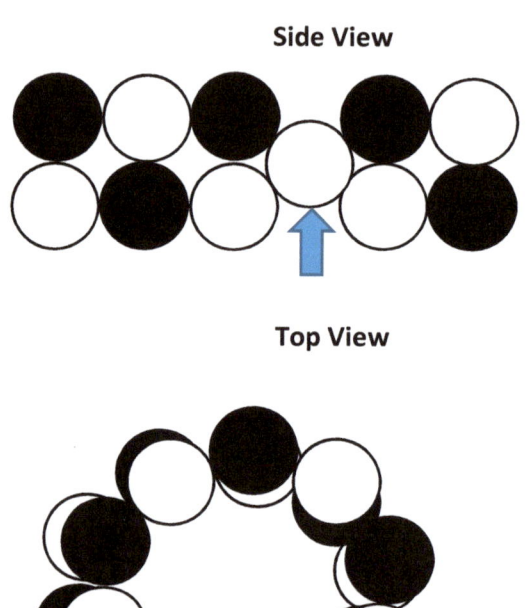

Top View

The spacer is important because otherwise, one of the layers might have one end with a proton touching the other end also with a proton.

Evidence:

Look the preferred isotopes of 026-Fe Iron. In nature, it seems it is never built perfectly 26 protons and 26 neutrons = 52 particles. Instead you get frequency of the isotopes of:

026-Fe Iron Isotopes	Frequency
53	0.05%
54	5.90%
55	0.05%
56	91.72%
57	1.20%
58	0.28
Total	100.00%

The most preferred for an even number of protons is also even. Even more preference in multi-rings even elements is that isotope has a multiple of four in total nucleus particles. My interpretation is that the best structure is two layers of twenty-eight (2 x 28 = 56) or four layers (4 x 14 = 56). The other favored isotopes are probably (2 x 27 = 54). But the true odd ones are not often chosen.

Something different happens when the number of protons start out as odd. Remember that the extra proton can sit at the end of a protrusion or to make the rows match a single neutron can separate and lock two odd-number rows in ring.

Side View

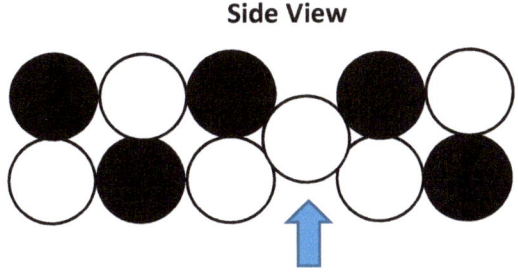

There must be a full substitution to get the rings the same. That means an extra neutron holding two rows, so odd. Or extra neutrons with that proton in a protrusion.

027-Co Cobalt Isotopes	Frequency
57	0.00%
58	0.00%
59	100.00%
60	0.00%
Total	100.00%

This separation of the neutrons means that only one path works to
build a larger nucleus. You must add the neutron first. That creates
two neutrons together. Then a proton can attach between those two
neutrons safely. Eventually, a path to a higher elements (more
neutrons) occurs, but only in that order. Otherwise, protons (+) touch
and nuclear decay of the structure.

A chain (segment of nucleus particles) is naturally proton-neutron-
proton and so on.

A neutron can attach to that segment.

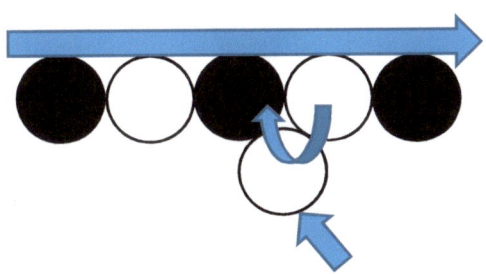

However, that structure creates an extra nucleomagnetics loop
besides the straight north-south direction of the nucleomagnetics
chain. This loop is a force holding that new neutron as a protrusion.

However, if a force works to break that nucleomagnetics loop, then the particles realign so a neutron-neutron block occurs.

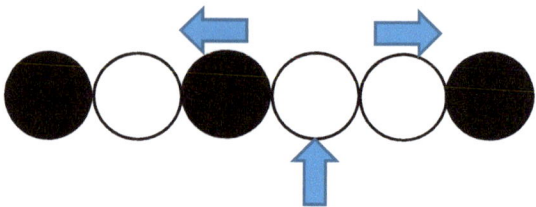

Only after that realignment, a proton can attach. Of course, the nucleomagnetics field for this intermediate stage also has a strange loop. Remember that must apply force in order of strength:

1) Charge is strongest. In this case, proton-proton repulsion

2) Magnetic is next. We must break the extra nucleomagnetics look before there is room for the next neutron.

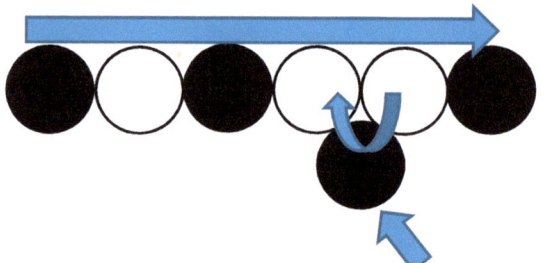

Of course, those can then re-orient back into a longer chain (and new element).

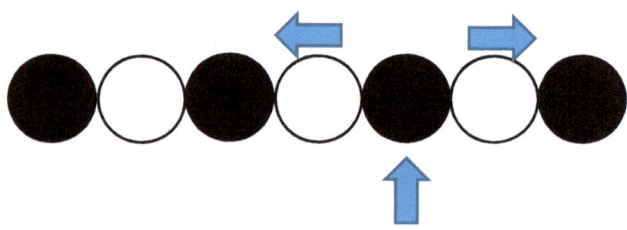

We see this in the radioactive element 088-Ra Radium. The structure is too big that a proton will eventually touch or near-touch each other and the element will change:

The number of protons is conserved:

Element Before	Elements After
088-Ra Radium	086-Rn Radon
	002-He Helium
Total # of Protons = 88	Total # of Protons = 88

The total number of nucleus particles is also conserved:

Element Before	Elements After
088-Ra Radium (isotope 226)	086-Rn Radon (222 isotope)
	002-He Helium (4 isotope)
Total # of Nuclear particles = 226	Total # of Nuclear particles = 226

Why the extra Neutron in some elements?

In the most common isotopes of these elements, some configuration like an extra neutron while others do not tend to contain the extra neutron:

Element	Atomic Number	Protons	# of Neutrons
003-Li Lithium	6.941	3	4
004-Be Beryllium	9.0122	4	5
005-B Boron	10.811	5	6
006-C Carbon	12.0107	6	6

Many times, the structure works with a standard ring, P-N-P-N- until back to the beginning. However, some structures work better with an extra neutron to create stability. 006-Carbon is a perfect 2:1 with twelve (12) nucleus particles for six (6) protons. For 006-C Carbon, we find N = P, so #particles = 2P or 2.000. Yet, the other elements are N = P+1.

Take for instance, Beryllium, with four (4) protons and the most common isotope with five (5) neutrons. This should only need four (4) neutrons to separate the four (4) protons into a ring. However, what happens more often is that an extra neutron arrive and one end of the ring become fixed. A ring eight around is extremely flexible. However, adding another neutron, and you get a box-and-diamond combination that is much less flexible.

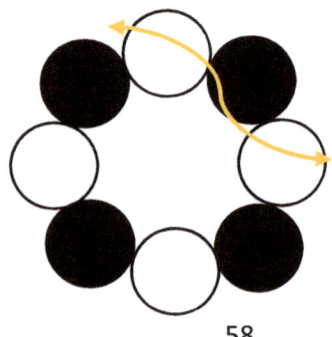

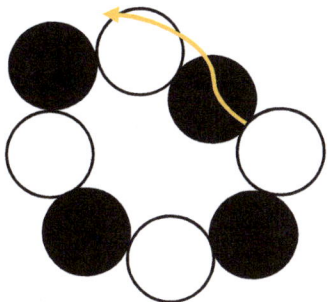

The above has lots of flexibility, maybe even too much so the protons would get too near each other. The below is very rigid with an extra neutron added. The added neutron makes the ring not able to twist. The nucleomagnetics double connection via the extra neutron makes this structure rigid.

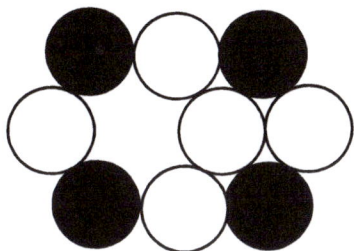

Therefore, you get a tendency for the 004-Beryllium to have five (5) neutrons. That makes the right end very rigid, and protons from the top do not incidentally get too close to a proton on the bottom right, as with the wobbly even-count structure.

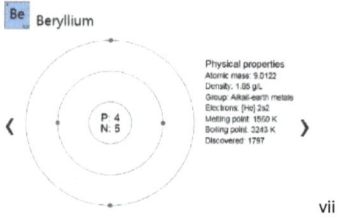

Beryllium

Physical properties
Atomic mass: 9.0122
Density: 1.85 g/L
Group: Alkali-earth metals
Electrons: [He] 2s2
Melting point: 1560 K
Boiling point: 3243 K
Discovered: 1797

P: 4
N: 5

vii

This give 004-Berylium a square proton configuration in the nucleus.

Why are certain large atoms radioactive? And why consistent exponential decay?

Because at large numbers and size, the combination with neutrons separating protons eventually have too many chains attached, and at a consistent rate, the chains touch another proton which makes them separate (decay) into two particles. The perfect rings I describe of small number of particle atoms are not common for the largest, radioactive atoms.

The large atoms are mixes of multi-layer rings with chains attached.

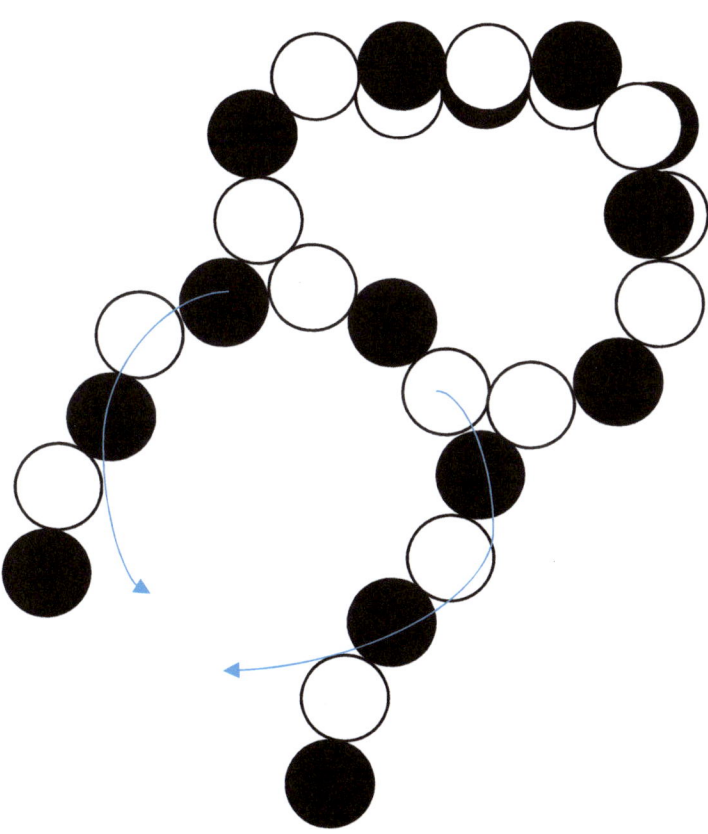

And, it two of chains get long enough and flop into each other, and that is radioactive decay.

The ratio is consistent, and as they meet and decay, the number goes down, exactly as per the radioactive decays charts.

Visualization of Nucleus Chained/Ring Configurations for Various Elements

Here are pictures (using small nucleomagnetics balls to visually simulate with red as the proton and green as the neutron) each element and its isoform (ring-configuration):

Element	Neutrons	Structure Form	Picture
H = Hydrogen	0	No ring – N/S	
001-H = Deuterium	1	No ring – N/S	
002-He Helium	2	Single ring	

Element	Neutrons	Structure Form	Picture
003-Li Lithium	3	Single ring	
004-Be = Beryllium	4	Single ring	
004-Be = Beryllium (Isotope 4 neutrons = 8 Atomic Weight)	5	Single ring	
004-Be = Beryllium (Isotope 5 neutrons = 9 Atomic Weight)	5	Locked ring	

Element	Neutrons	Structure Form	Picture
005-B = Boron (Isotope 6 neutrons = 11 Atomic Weight)	6	Single ring	
006-C = Carbon (coal)	6	Single ring	
006-C = Carbon (graphite)		Linked Rings	
006-C = Carbon (diamond)		2-layers of rings	

Element	Neutrons	Structure Form	Picture
007-N Nitrogen	8	2-layers of ring 1-proton sticking out	
008-O Oxygen	8	Single ring locked	
011-Na Sodium		2-layers of rings	
012-Mg Magnesium		2-layers of rings	

Until you have tried to make nucleomagnetics chains fit in a certain way, and kept the protons apart, you cannot understand the ways in which the loops of magnets can only build in certain ways.

Try it!

For example, you can try to create two lines:

Side View

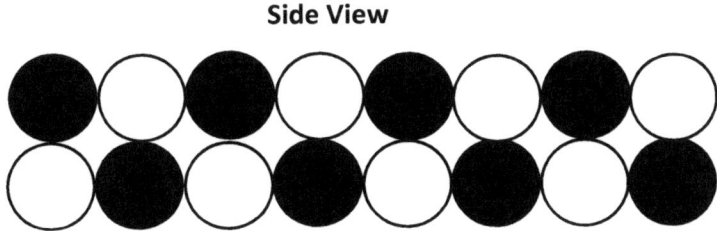

However, the nucleomagnetics forces cannot rotate at the ends. The magnet will magically move back to loops at the end.

Top View

For the advanced explorer, you can calculate why those ends cannot make the 90, then 90 degree turns.

For the people that learn by experience, try it!

Multi-Layer Rings Based upon Number of Particles (by Element)

As I showed, the nucleus can arrange in any way. However, from the nucleomagnetics moment calculation, you can guess that certain configurations are most common based upon the number of particles.

Element	Protons	Neutrons	Structure Form
001-H - Hydrogen	1	0	1 particle
001-H - Deuterium	1	1	chain – N/S
001-H – Trillium	1	2	chain – N/S
002-He – Helium	2	2	Single ring
Li – Lithium	3	3	Single ring
Be – Beryllium	4	4	Single ring
B – Boron	5	5	Single ring
C - Carbon (coal)	6	6	Single ring
C - Carbon (graphite)	6	6	2 flat rings - Infinity-8 loop
C = Carbon (diamond)	6	6	Double ring
C13 - Carbon+1Neutron (coal)	6	7	2 flat rings - Infinity-8 loop
C14 - Carbon+2Neutrons (coal)	6	8	2 flat rings - Infinity-8 loop with extras locking in center
N – Nitrogen	7	8	Double ring
O – Oxygen	8	8	Double ring
F – Florine	9	10	Double ring
Ne – Neon	10	10	Double ring
Na – Sodium	11	11	Double ring

Mg – Magnesium	12	12	Double ring
Al – Aluminum	13	13	Double ring
Si – Silicon	14	14	Double ring
Ph – Phosphorus	15	15	Double ring
S – Sulphur	16	16	Triple ring
Cl – Chlorine	17	17	Triple ring
Ar – Argon	18	18	Triple ring
K – Potassium	19	19	Triple ring
Ca – Calcium	20	20	Triple ring

Isotopes and Isoforms

For a century, there have been a concept of isotopes. That is a classification where the nucleus with the same number of protons (same Element), gets grouped (as isotopes) based upon the number of neutrons.

However, the proposed structure allows nucleus to vary in other ways. You can have nuclei with the same grouping of one number of protons and a certain number (for that isotope) of neutrons, as such the same isotope, but the physical structure is different.

002-He Helium (4 Isotope – chain isoform)

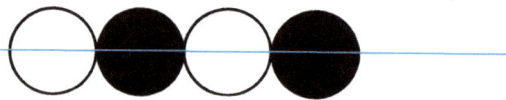

002-He Helium (4 Isotope – 1-ring isoform)

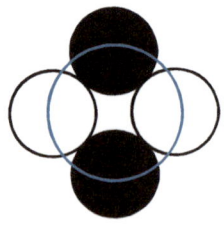

Both of these atoms have 2 protons and 2 neutrons = 4 nucleus particles.

However, the nucleomagnetics of the two are very different. The 1-ring has lower strength, and a nucleomagnetics pole orientation at 90 degrees to the ring. However, the chain isoform has stronger nucleomagnetics and a nucleomagnetics orientation directly in line with the nucleomagnetics of the individual particles.

Further, the linear isoform remains very flexible. The 1-ring structure is very rigid.

The same happens with isotopes and isoforms of 006-C Carbon (C-14 Isotope – 1-ring isoform extra neutrons outside) is flexible.

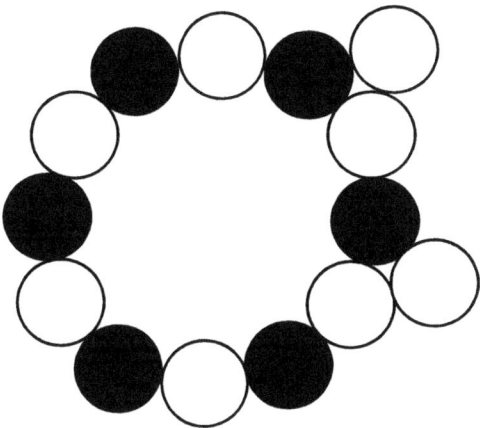

But the same isotope is more rigid if the extra neutrons bind inside.

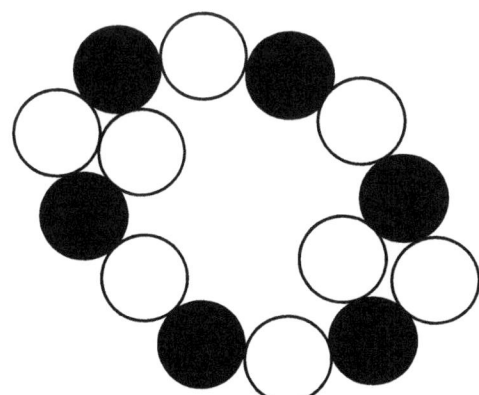

Proposal: Transparency and the Nucleus Structure – 006-C Carbon as Coal versus Diamond

In case you missed, I showed three isoforms of 006-C Carbon.

006-C = Carbon (coal)			
006-C = Carbon (graphite)			
006-C = Carbon (diamond)			

If the structure of the nucleus is flexible, then it can absorb energy easily. Lots of frequencies can move the particles, and thereby do not move on.

However, a 2-ring structure is more rigid.

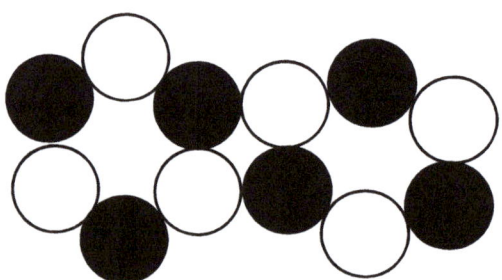

However, a 2-layers-of-rings structure is extremely rigid. Nothing can move.

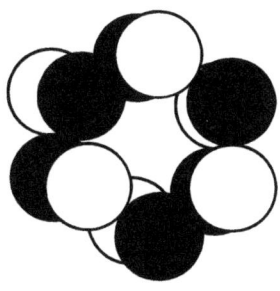

Isoform	Flexibility	Range of frequencies absorbed	Transparency
1-ring	Lots	Up to	None (black)
Flat 2-ring	Limited	Only small range	Silvery
2-layers of rings	None	Almost none	Transparent

The color of these absorption or transparency is based upon the diameter of a nucleus.

There is another source of color absorption from the structure of the crystalized bonding. If the electron bonds are oriented in a crystal, that covers absorption or translucency of frequencies in the range of that distance. That limits the frequencies by this locking.

My postulate is that you have both types going on in clear crystals.

However, the clearness, clarity, to see through you need all frequencies limited in absorption, at both nucleus, electron shell, and bonding.

Nucleomagnetics Binding Creates
Preferred Angles and Structures

Magnetics has come amazing quirks. As already stated, it does not fit in neat rows. The magnetic field of one row will create pressure on the next row. Here is one example.

Top View

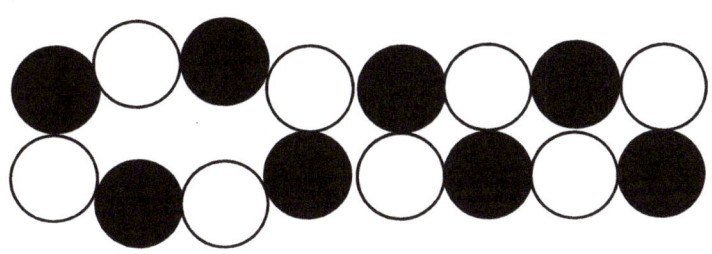

Top View

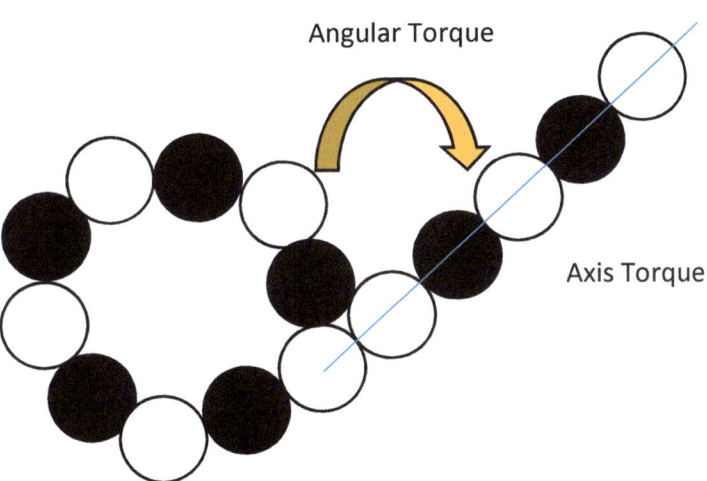

You will get a great understanding by fiddling with two colors of ball magnetics to really understand. Some things work. Some configurations do not work. Magnetics at close distances is very, very powerful.

Challenge: Observation Not Currently Possible within Nucleus

No one can observe the actual workings at this distance. We guess and interpret based upon deflecting rays, based upon the force on nearby particles, and other means.

Nucleus Internal Structure per the AVSC Arno Vigen Scrunched Cube Atomic Model

Let's go through sections of the Periodic Chart that build differently.

The first

Element # / Name	If 2:1	Atomic Mass	Change from Last	Sets of 2
01-H-Hydrogen	2	1.01	1.01	
02-He-Helium	4	4.00	2.99	4.00
03-Li-Lithium	6	6.94	2.94	
04-Be-Beryllium	8	9.01	2.07	5.01
05-B-Boron	10	10.81	1.80	
06-C-Carbon	12	12.01	1.20	3.00
07-N-Nitrogen	14	14.01	2.00	
08-O-Oxygen	16	16.00	1.99	3.99

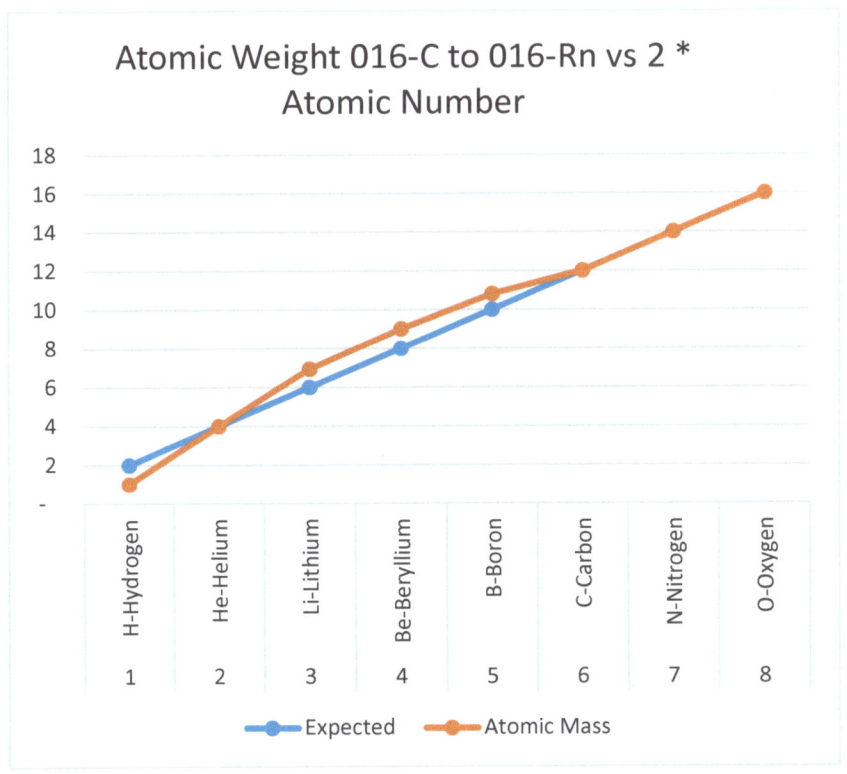

Atomic Weight 016-C to 016-Rn vs 2 * Atomic Number

Hydrogen, being 1, does not required the proton-neutron pairing. 002-He-Helium, 006-Carbon, 007-N-Nitron, and 008-O-Oxygen are all perfect fits. The structure is proton-neutron-proton-neutron.

In AVSC, these are single-rings.

Next, Elements build where to fit two more protons, you need two neutrons, a direct proton-neutron-proton-neutron structure. However, there are bumps.

In AVSC Scrunched Cube Atomic Model, these are double-rings. The extra on the odd is the single-separator so the two layers can grow neutron-single first.

Element # / Name	Expected	Atomic Mass	Change from Last	Sets of 2
06-C-Carbon	12	12.01		
07-N-Nitrogen	14	14.00	1.99	
08-O-Oxygen	16	16.00	2.00	3.99
09-F-Florine	18	19.00	3.00	
10-Ne-Neon	20	20.18	1.18	4.18
11-Na-Sodium	22	22.99	2.81	
12-Mg-Magnesium	24	24.31	1.32	4.13
13-Al-Aluminum	26	26.98	2.67	
14-Si-Silicon	28	28.09	1.11	3.78
15-P-Phosphorus	30	30.97	2.88	
16-S-Sulfur	32	32.07	1.10	3.98

Visually, this is a clear pattern. Every two protons gets two neutrons. Further, all the odds count Elements has an extra one because you cannot build the first proton increase without an electron on each side to separate the perfect alignment that already exists in the even-count protons.

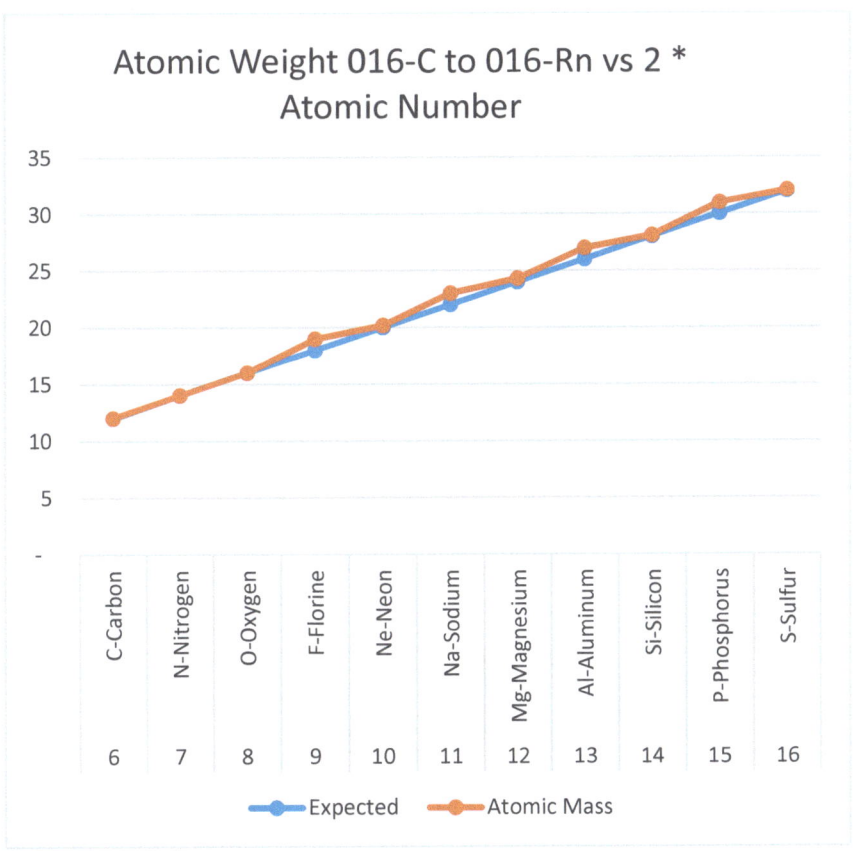

Atomic Weight 016-C to 016-Rn vs 2 * Atomic Number

You can observe that for this nucleus physical structure, the Atomic Weight increases by three (3), then by one (1), then by three (3), and then by one (1) again. It is a clear pattern.

In my AVSC Scrunched Cube Atomic Model, this is 'three-then-one' building is a double ring. The neutron must get added first, and in both directions or rings. The double ring gets a proton-neutron pair, but that would make the ends meet proton-proton, so the third particle, the neutron separator must occur for odd numbered double-ring nucleus.

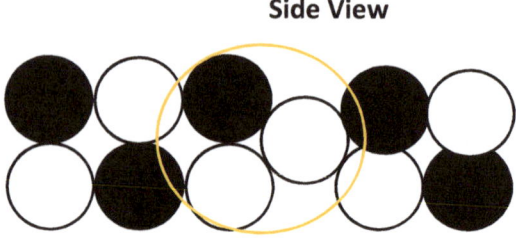

It must be a set of three.

Of course, to get from that to an even-count Atomic Weight, you only need a proton from the correct direction.

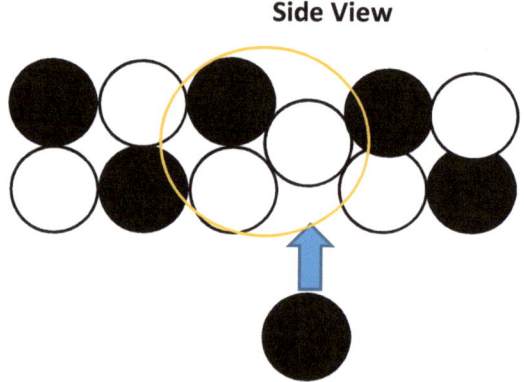

So, it becomes alternating proton-neutron, with a 2nd layer neutron-proton. It is very stable.

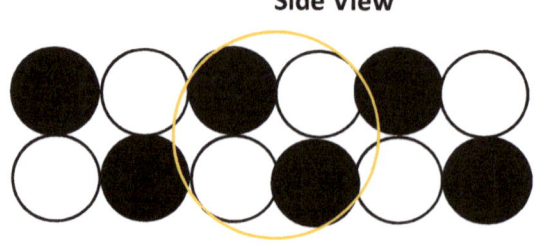

That is how from even to even becomes three (3), then one (1) with two neutrons needed to add two protons to a nucleus.

Yet something changes in the next section. Eventually, the single ring gets too big, and wobbly. Then, the double ring gets too big, and wobbly. Based upon this the number around a ring before it gets to big is:

Atomic Number	Symbol- Element Name	Expected	Atomic Mass	Change from Last	Sets of 2
11	Na-Sodium	22	22.99	2.81	
12	Mg-Magnesium	24	24.31	1.32	4.13
13	Al-Aluminum	26	26.98	2.67	
14	Si-Silicon	28	28.09	1.11	3.78
15	P-Phosphorus	30	30.97	2.88	
16	S-Sulfur	32	32.07	1.10	3.98
17	Cl-Chlorine	34	35.45	3.38	
18	Ar-Argon	36	39.95	4.50	7.88
19	K-Potassium	38	39.10	(0.85)	
20	Ca-Calcium	40	40.08	0.98	0.13
21	Sc-Scandium	42	44.96	4.88	
22	Ti-Titanium	44	47.88	2.92	7.80
23	V-Vanadium	46	50.94	3.06	
24	Cr-Chromium	48	52.00	1.06	4.12

The structure works 2:1 all the way to 16-S-Sulfur. Perfect to 32 particles. 2 * 2 * 2 * 2 * 2. Two rows of 16, or 4 rows of 8 around.

Yet, something happens at 18-Ar-Argon with 36 that makes the structure jump, then bind protons without adding neutrons for the next few Elements. Yet is back to perfect at 20-Ca-Calcium again.

Now, 36 is one of those magic numbers. It is 2 * 2 * 3 * 3. It could be the nice limit going from 2 * 18 = 36 particles. Or it could be 3 * 12 = 36. Or, it could be 4 * 9 = 36 particles.

Something happens at 21-Sc-Scantium that makes the next section use up eight (8) particles to add two protons.

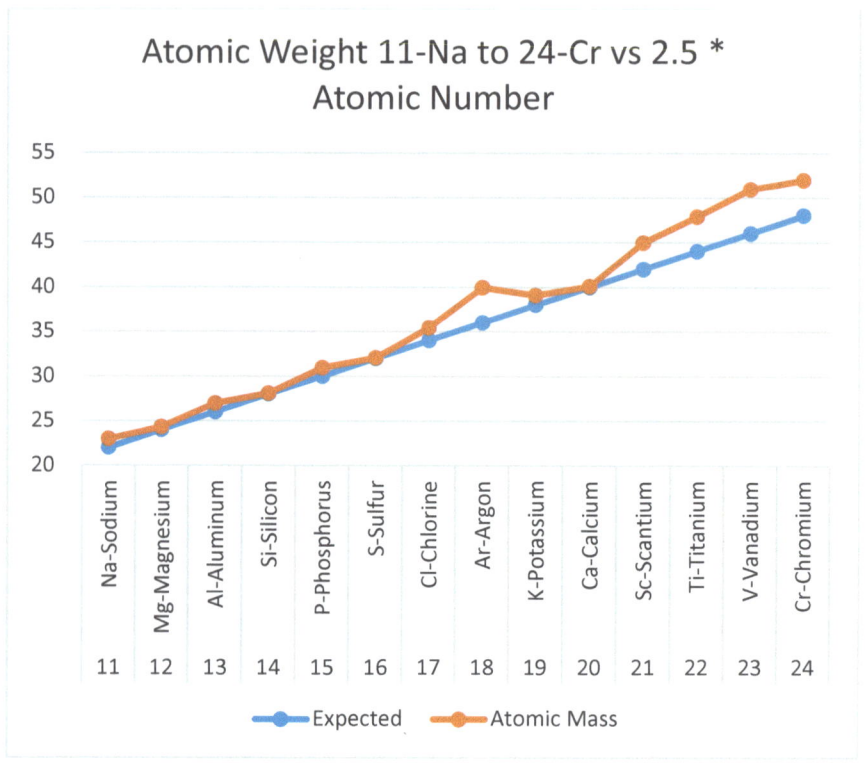

There is a strange bump, then a permanent separation where the 2.0:1 becomes 2n + 4 for even-proton-count Elements and 2n+5 for odd-proton-count Elements. Therefore, every other layer built needs two neutrons to separate the ring layer. That special connector section is not 2:0, but 6 neutrons to connect 2 or 3 protons.

To get to a wider cylinder of nucleus particles, the structure cannot build as easily. Every so often, an extra neutron set of spacers occurs. If we have:

	N		P		N

Meeting

	N		P		N

The only want that works is by:

	N		P		N
		N		N	
	N		P		N

There is six-neutrons to connect two protons. Four extra neutrons are necessary to handle the end to end match.

Side View

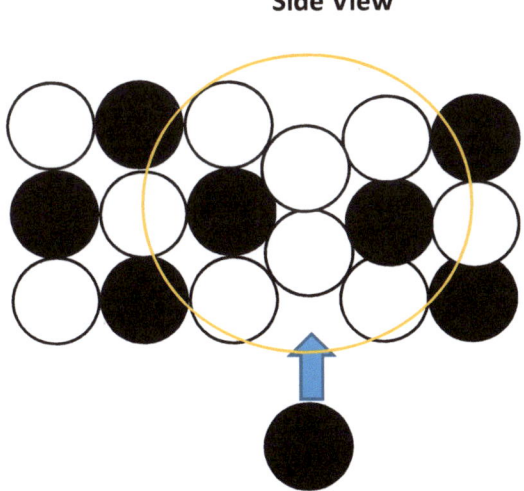

However, when that gets another proton, it can only be from one side.

Side View

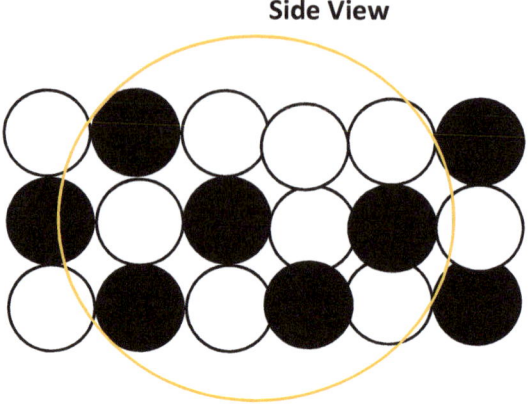

So, a well-built set of even, then odd, section of the nucleus of a triple-ring nucleus has going left to right:

P	N	P
N	P	N
P	N	N
N	P	N

That is twelve particles, five protons and seven electrons. The ratio of particles (Atomic Weight) to protons (Atomic Number) is 2.4:1, not the 2.0:1 of the single-ring and double-ring nucleus structures.

Half Offset makes rows very, very stable

The other reason for this layer that you don't get dangerous shift.

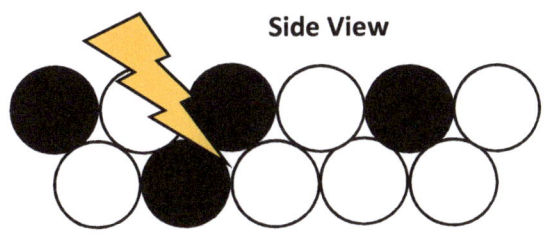

Yet, with the extra section of a) neutrons only and b) one less than the row, so that a collapse cannot occur.

Side View

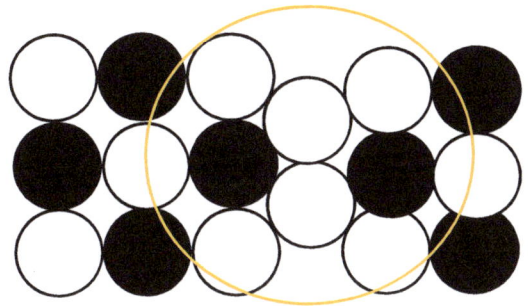

Again, I suggest that every student tries this with a set of magnetic balls. You can learn what can work and cannot work given the force of magnetics. Remember:

Nucleomagnetics is the most powerful force inside the Bohr radius.

A similar process can occur in another building direction; the cylinder can get thicker.

Side View

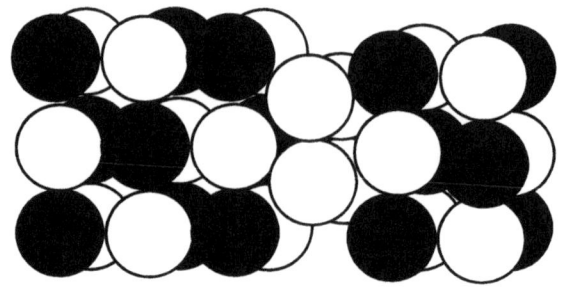

Around the ring, a nucleus then has a cylinder, empty in the middle.

3 wide x 2 deep = 6

3 wide x 3 deep = 9

This intermediate structure creates the method to get from 36 to 72 particles.

This leads to a section of where the Atomic Weight (protons plus neutrons) increase versus the Atomic Number (protons) at a new rate of 2.5:1.

Symbol-Element Name	Expected	Atomic Mass	Change from Last	Sets of 2
24-Cr-Chromium	53	52.00		
25-Mn-Manganese	55	54.94	2.94	
26-Fe-Iron	58	55.85	0.91	3.85
27-Co-Cobalt	60	58.93	3.09	
28-Nickel	63	58.69	(0.24)	2.85
29-Cu-Copper	65	63.55	4.85	
30-Zn-Zinc	68	65.38	1.83	6.69
31-Ga-Gallium	70	69.72	4.34	
32-Ge-Germanium	73	72.63	2.91	7.25
33-As-Arsenic	75	74.92	2.29	

34-Se-Selenium	78	78.97	4.05	6.34
35-Br-Bromine	80	79.90	7.27	
36-Kr-Krypton	83	83.80	3.90	11.17
37-Rb-Rubidium	85	85.47	6.50	
38-Sr-Strontium	88	87.62	2.15	8.65

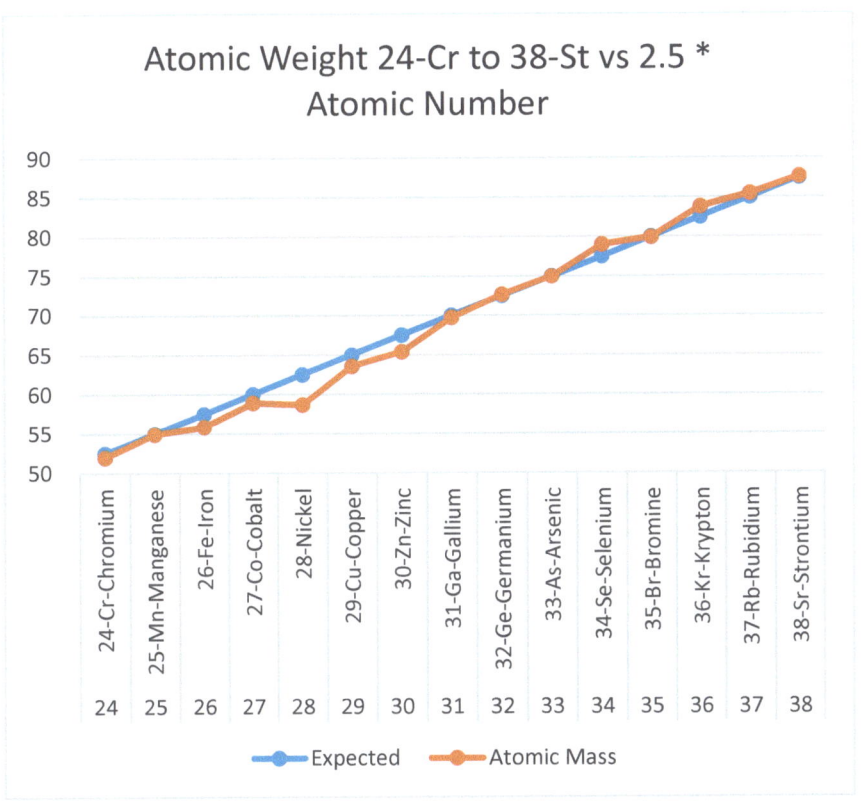

Yet, some of the original structure also needs extra neutrons, to fit into the 3D cylinder. Therefore, extra neutrons are needed until the entire structure is at particles:protons of 2.5:1.

This continues to a range where the additional are clearly 5 particles for every 2 protons (2.5:1):

Symbol-Element Name	Expected	Atomic Mass	Change from Last	Sets of 2
38-Sr-Strontium	88	87.62		
39-Y-Yttrium	91	88.91	1.29	
40-Zr-Zirconium	93	91.22	2.32	3.60
41-Nb-Niobium	96	92.91	1.68	
42-Mo-Molybdenum	98	95.95	3.04	4.73
43-Tc-Technetium *	101	**98.00**	2.05	
R	103	101.07	3.07	5.12
45-Rh-Rhodium	106	102.91	1.84	
46-Pd-Palladium	108	106.42	3.52	5.35
47-Ag-Silver	111	107.87	1.45	
48-Cd-Cadmium	113	112.41	4.55	5.99

One interesting note is that the section at the end of the low increase is radioactive. The postulate that a full layer (width or depth of rings) gets to a limit. It tries to keep that smaller structure, but that leaves a stray protons exterior at a new layer alone. As a result, it is easily dislodged, and radioactive decay back to the stable.

This happened at 18-Ar-Argon with a strange bump at the end of the perfect 2.0:1 range. With 36 particles, it is a great range of 2*2*3*3 which means the structure can be 4 x 9, 6 x 6 (my guess), or 2 x 18.

Here the radioactive, linked to a position that change from at low ratio section (even-to-even < 4) moving to a clear (even-to-even >5 and even 6). The Element at that transition has trouble staying together.

We will re-visit radioactivity much deeper. For large atoms, it really gets really interesting.

The interesting observation about this transition point is that it occurs at 100 particles, which is 2*2*5*5, which could be 4*5*5 a well-balanced limit. Or it could be 108 which is 2*2*3*3*3 less the open core section of 9 for 99 particles as the structure limit.

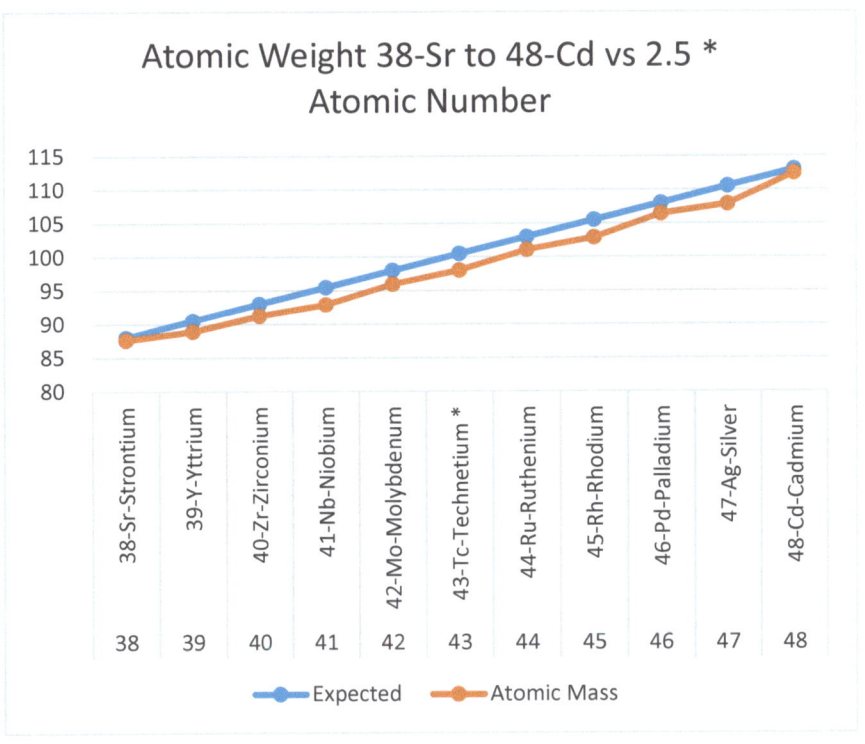

* 43-Tc-Technetium is radioactive.

The growth continues at 2.5:1 which includes another radioactive point at 144 particles. Notice that 144 = 2*2*3*3. Alternatively, you can this of that as 2*3*5*5 = 150 less that core of 5 particles for 145.

In either case, there is a sudden slow in the particle growth ending with a radioactive structure, then the section after returns to a 2.5:1

growth rate from the base particle count (with those previous structure as the base).

* 61-Pm-Promethium is radioactive.

Symbol-Element Name	Expected	Atomic Mass	Change from Last	Sets of 2
49-In-Indium	117	114.82		
50-Sn-Tin	119	118.71	3.89	
51-Sb-Antimony	122	121.76	3.05	6.94
52-Te-Tellurium	124	127.60	5.84	
53-I-Iodine	127	126.90	(0.70)	5.14
54-Xe-Xenon	129	131.29	4.39	
55-Cs-Cesium	132	132.91	1.61	6.00
56-Ba-Barium	134	137.33	4.42	
57-La-Lanthanum	137	138.91	1.58	6.00
58-Ce-Cerium	139	140.12	1.21	
59-Pr-Praseodymium	142	140.91	0.79	2.00
60-Nd-Neodymium	144	144.24	5.34	
61-Pm-Promethium *	147	**145.00**	0.76	6.10
62-Sm-Samarium	149	150.36	9.45	
63-Eu-Europium	152	151.96	1.60	11.06

In the below, you can see the bulge above the line, then flatting and the radioactive element as the turning point. Further, the radioactive elements get found at point below a certain ratio.

It is especially pronounced near these 2 and 3 combinations. In quantum mechanics, these are fancy step functions where 'color' determines when it is functionality changes at a 2, and when a 3 works.

In AVSC, it is simply the famous map challenge, to draw a map of protons where the same type cannot touch. The great challenge is building the no-touch protons is a 3D map.

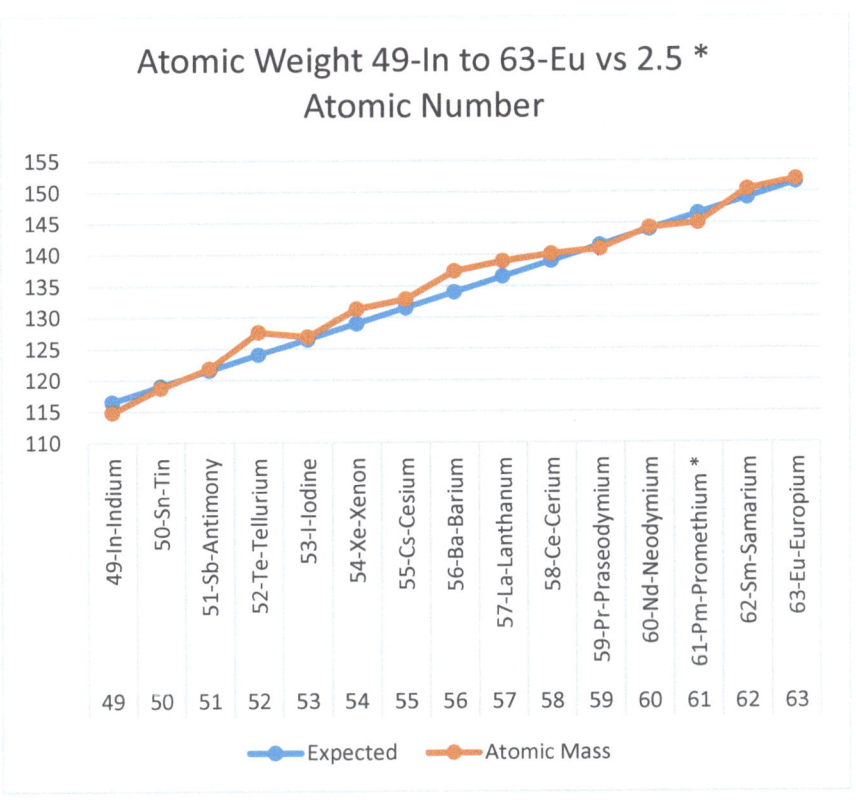

This is a stable process for this range of elements. However, the change place is at 61-Promethium. Notices that 61- is not where electrons shells change

In AVSC, electron shells get full and start new shells in spherical 3D hemispheres.

In AVSC, nucleus layers change in cylinder width and depth with an open core.

The full layers in nucleus structure has no relation to the full shells of electrons.

Finally, there begins the section where the number of neutrons required to bind a proton without touching another proton becomes even greater (11:1).

In AVSC, this is the depth/thickness of the cylinder becoming a third layer. That is the structure becomes 3 layers, open center, 3 layers. That will become important as the final stable fully-loaded structure of a nucleus gets created at 216 less 6 for 210 particles.

Symbol-Element Name	Expected	Atomic Mass	Change from Last	Sets of 2
64-Gd-Gadolinium	157	157.25		
65-Tb-Terbium	160	158.93	1.68	
66-Dy-Dysprosium	162	162.50	3.57	5.25
67-Ho-Holmium	165	164.93	2.43	
68-Er-Erbium	167	167.26	2.33	4.76
69-Tm-Thulium	170	168.93	1.68	
70-Yb-Ytterbium	172	173.05	4.12	5.80
71-Lu-Lutetium	175	174.97	1.91	
72-Hf-Hafnium	177	178.49	3.52	5.44
73-Ta-Tantalum	180	180.95	2.46	
74-W-Tungsten	182	183.84	2.89	5.35
75-Re-Rhenium	185	186.21	7.72	
76-Os-Osmium	187	190.23	4.02	11.74
77-Ir-Iridium	190	192.22	8.38	
78-Pt-Platinum	192	195.08	2.87	11.24

You can see the graph of this range begin to race upwards after 74-W-Tunsten with 184 particles. This number being 192 (2*2*2*2*2*3) less 8 for the core.

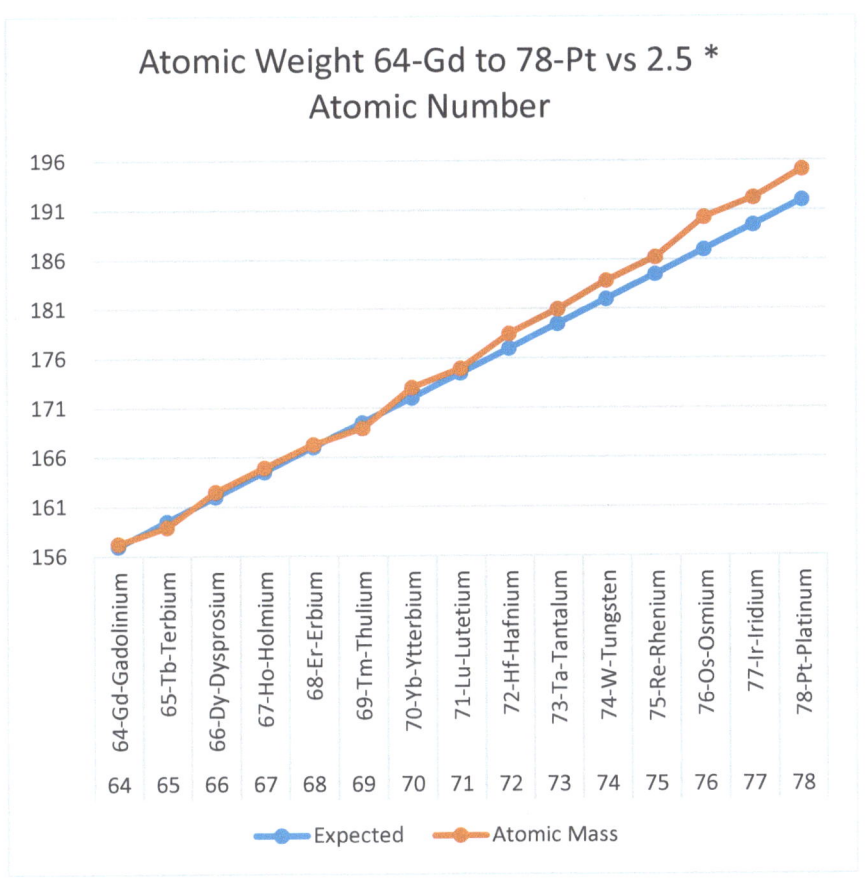

Central Channel of Nucleus Remains Empty along its induced nucleomagnetics axis because avoiding protons as the center is almost impossible.

A position along the nucleomagnetics axis touches two positions along the nucleomagnetics axis, and another six position around the cylinder. A proton cannot avoid eight positions that might have protons. The natural solution becomes to build structures where the axis row is empty.

What are the Structures of Protons and Neutrons as Build the Nucleus of Larger Molecules?

Because of the nature of these proton building, we see different types of jumps between the most stable isotopes of each Elements. These changes are based upon the size and shape of the base nucleus proton-neutron configuration.

Ratio: 2.0:1
Group Jumping by four (4) particles – two (2) protons plus two (2) neutrons creates stable growth:

Group Jumping by four (5) particles – two (2) protons plus two (2) neutrons creates stable growth:
Ratio: 2.5:1

1) Even to odd – jumping by three (3) for the extra neutron at the end bridging two layers

2) Odd to even – jumping by one (1) making the two

   Group Jumping by more than five (5) particles – two (2) protons plus two (2) neutrons creates stable growth:
   Ratio: 2.5:1

3) Even to odd – jumping by three (3) for the extra neutron at the end bridging two layers

4) Odd to even – jumping by one (1) making the two

Ratio Jumps to >2.5 for Large Radioactive Elements

The nucleus remain stable to 210 particles. In fact, the last four elements – 78-Pb-Lead, 83-Bi-Bismuth, 84-Po-Polonium, and 85-At-Astatine – seem to level off at 207, 208, 209, then 210 particles. However, at the start of radioactivity, the next elements rquires 222 nucleus particles for some stability (not completely because it is radiactive). That is a lot of neutrons that need to get added before next proton can fit in.

Neutrons Added = 12

Proton Added = 1

Finally, the last stable group and the start of the full radioactive heavy elements.

97

Symbol-Element Name	Expected	Atomic Mass	Change from Last	Sets of 2
76-Os-Osmium	190	190.23		
77-Ir-Iridum	193	192.22	1.99	
78-Pt-Platinum	195	195.08	2.86	4.85
79-Au-Gold	198	196.97	1.89	
80-Hg-Mercury	200	200.59	3.62	5.51
81-Tl-Thallium	203	204.59	4.00	
82-Pb-Lead	205	207.20	2.61	6.61
83-Bi-Bismuth	208	208.98	1.78	
84-Po-Polonium	210	**209.00**	0.02	1.80
85-At-Astatine	213	**210.00**	1.00	
86-Rn-Radon	215	**222.00**	12.00	13.00
87-Fr-Francium	218	**223.00**	1.00	
88-Ra-Radium	220	**226.00**	3.00	4.00

In a graph, you can see the leveling off at 210, and the huge jump to 222 as the Element nucleus builds differently.

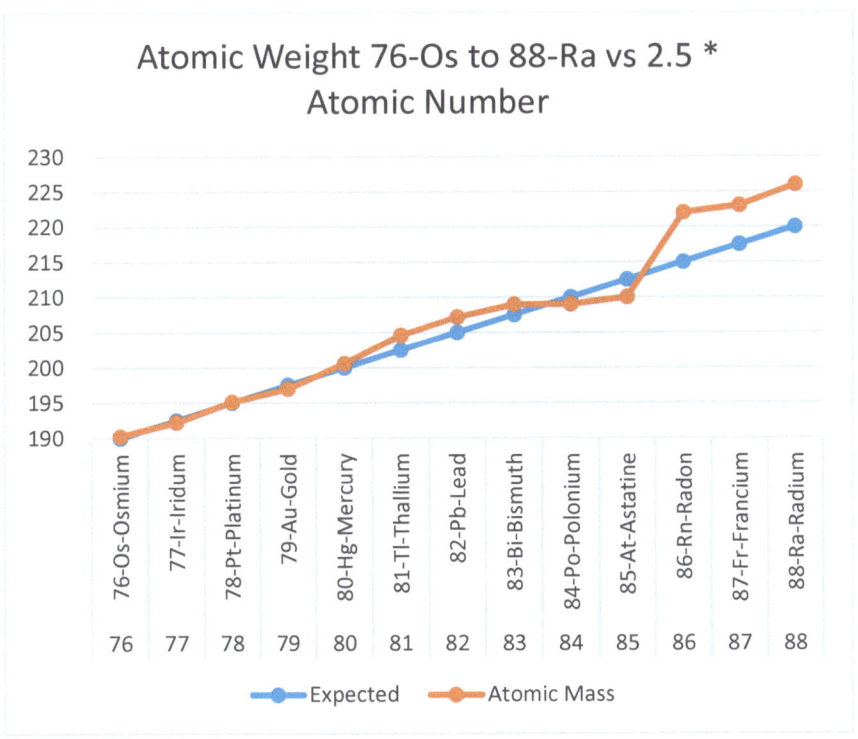

Atomic Weight 76-Os to 88-Ra vs 2.5 * Atomic Number

There it is that budge, then dip below the 2.5:0 ratio, and that is the radioactivity transition point.

In AVSC, the 3D particle layered structure is 216 less 6 = 210, and the 216 is 2*2*2*3*3*3 or 6*6*6. The very center six (6) touch to many and would build empty. That makes 210 the largest stable nucleus limited to patters of six.

Here is my idea of the nucleus structure. It is somewhat like a cylinder. It consists of an inner open six . The outer layers have to have extra neutrons to keep protons from touching the smaller inner layers that would hit more than one particle.

The inner layers are generally proton-neutron-proton-neutron-proton (P-N-P-N-P):

The outer layers would need extra neutrons where the inner layer is a proton so that pattern might be proton-neutron-neutron-neutron-proton (P-N-N-N-P):

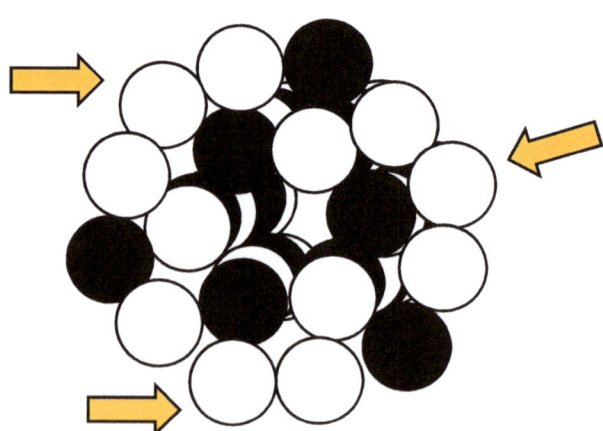

Where the outer layer has these extra neutrons, then the next layer can go back more towards P-N-P-N with a few exceptions.

One layer having a few proton (yellow arrows), which leads to the next layer only needing a few neutrons (blue arrows).

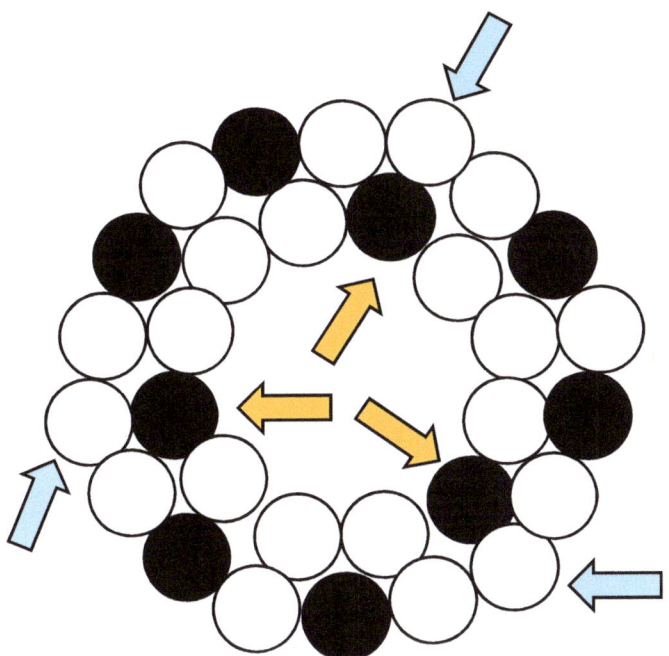

The patter goes as the layers settle between needing lots of neutrons to needing only a few neutrons.

That full 6x6x6 layer becomes in the place where adding a new layer, Layer 7, one new proton that does not hit the many neutrons in Layer 6, so it requires lots of neutrons.

Even more interesting, the extra might even become these flapping tails.

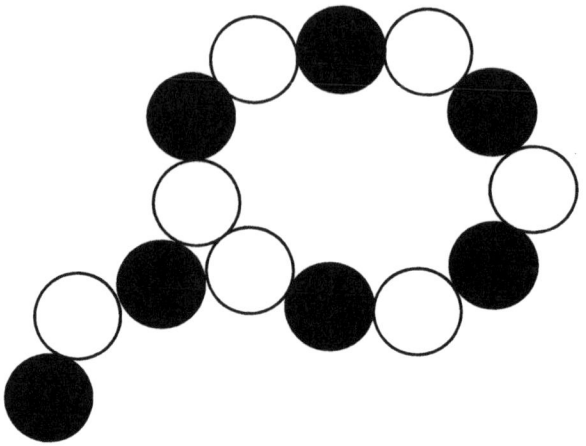

And, if that tail flaps to get too close to another proton, then
radioactive decay.

Why does a Nucleus Become Unstable and Decay?

The factors, in order of priority, for the decay rate of heavy radioactive nucleus include:

- Angle of approach of particles
- Number of particles approaching
- Number of particles ejected
- The structure of the outermost layer of the nucleus.

The angle of approach is critical; some direction settle in seven (7) layers of subshells position electrons, so thick that it block any approach to the nucleus. The nucleus axis always have a subshell at both nucleomagnetics axis poles; therefore, an approaching particles (proton or neutron) would need to get past the energy blockage of each of those seven layers. That is very unlikely.

AFSC teaches that electron subshells build from the nucleomagnetics axis towards the equator. That means that the equator area, for some Elements is very open.

The middle two factors are critical to radioactivity. For a material of decent size (more than a few molecules), the third factor, # Ejected, also becomes the second factor, # Ejected. In this way, the change to radioactivity if squared (exponential). Double (2x) the ejections and the decay rate increases fourfold (4x).

There is point after 082-Lead which has 210 Atomic Weight, that the nucleus of any atom of a higher Element is unstable, and decay with a mathematical certainty of rate.

Something magically happens at the Atomic Weight of 210. You can see something changes in the structure.

Symbol-Element Name	Expected	Atomic Mass	Change from Last	Sets of 2
83-Bi-Bismuth	208	208.98	1.78	
84-Po-Polonium	210	209.00	0.02	1.80
85-At-Astatine	213	210.00	1.00	
86-Rn-Radon	215	222.00	12.00	13.00
87-Fr-Francium	218	223.00	1.00	
88-Ra-Radium	220	226.00	3.00	4.00
89-Ac-Actinium	223	227.00	1.00	
90-Th-Thorium	225	231.04	4.04	5.04
91-Pa-Protactinium	228	231.04	-	
92-U-Uranium	230	238.03	6.99	6.99

Also, notice that after the transition point the next structure is 4 particles for 2 protons. It is a hanging chain just like the very small chain and single ring nucleus structures.

Of course, after that it gets more difficult to find a spot quickly with 5:1, then 6:1 ratios.

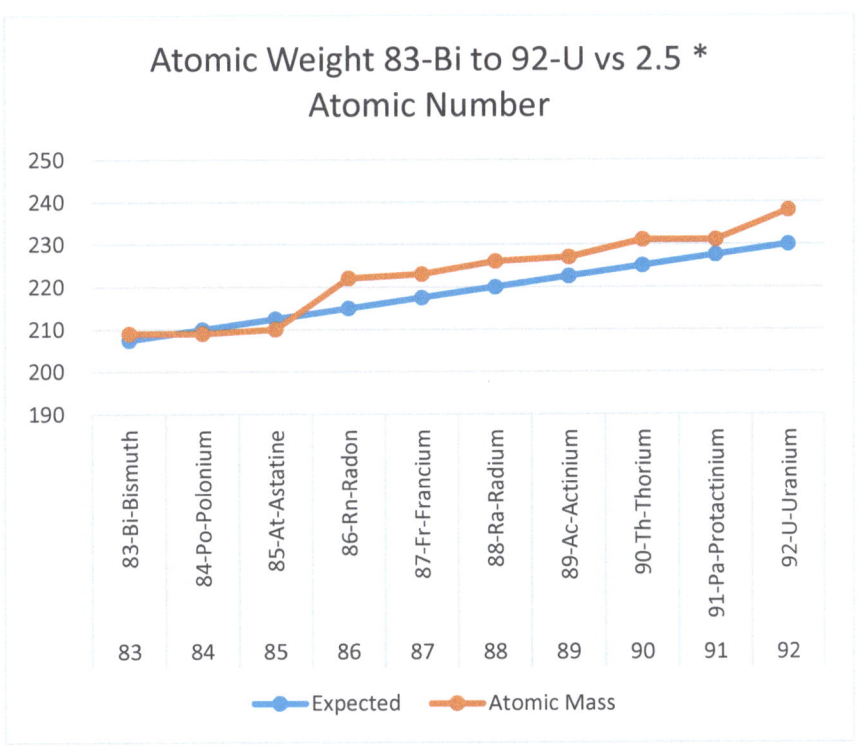

Atomic Weight 83-Bi to 92-U vs 2.5 * Atomic Number

At 086-Rn-Radon, elements become radioactive. Radioactive elements tend to decay back to an elements like 086-Pb-Lead. There is something fundamental happening at that number of protons, the Atomic Number, and Atomic Weight, the number of nucleus particles, the sum of protons and neutrons.

Yet, more important is the comparison of the nucleus decay rate relative to the excess particles count above 210, the last stable level. The comparison is based upon only the various Elements and isotopes in the same subshell-7t.

Element	Neutron Count	Excess	Ratio	Decay Rate
Th-Thorium	142.04	16.04	2.67	1.41E-07
Pa-Protactinium	140.04	14.04	2.01	9.20E-05
U-Uranium - average	146.03	20.03	2.50	2.49E-07
U-Uranium-Isotope-235	143.00	17.00	2.13	1.14E-05
U-Uranium-Isotope-237	145.00	19.00	2.38	1.87E-06

This may look like a lot of variety, but when reviewed as a scatter chart, a clear pattern exists.

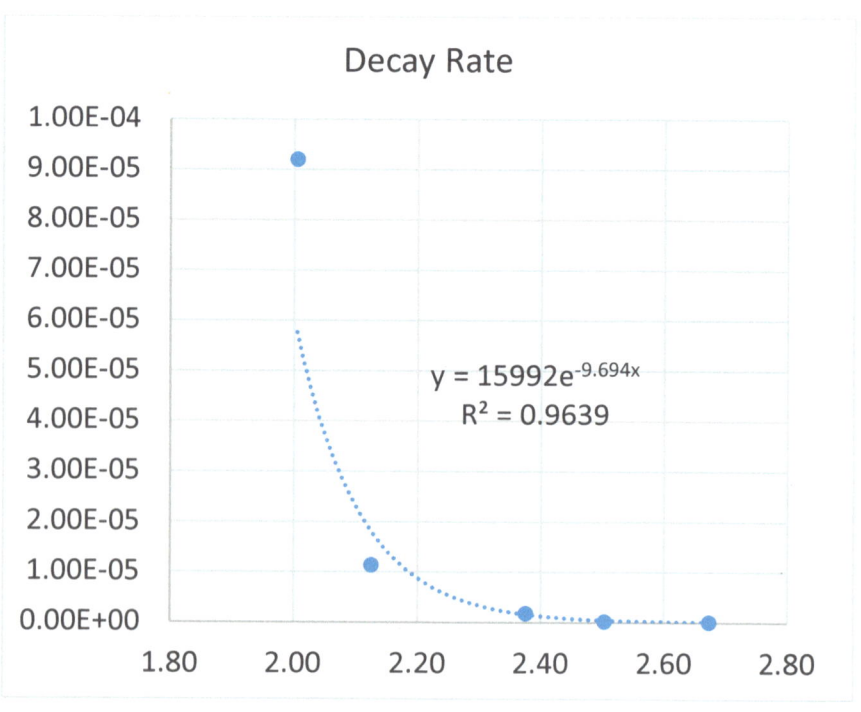

The exponential decay relative to that excess layer ratio of particles to protons expresses a 96% R-square correlation of prediction versus actual. That is very significant.

Review

In the Arno Vigen Scrunched Cube model, the nucleus is a set of rings with extras sometimes on the outside. It was easy to visualize for low-count Element, like Carbon with twelve (12) particles. It is a simple proton-neutron-proton-neutron as one long ring of twelve (12) or two-rings of six (6) with alternating proton-neutron on one, and neutron-proton on the other. That configuration keeps a neutron separating every proton.

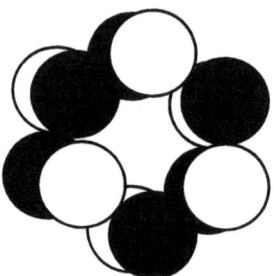

This logic work very well up to 2 around x 16 rings as alternating if that were built perfectly. Up to 016-S-Sulfur, the ratio of proton to neutron is 2:1.

Symbol-Element Name	Expected	Atomic Mass	Change from Last	Sets of 2
06-C-Carbon	12	12.01		
07-N-Nitrogen	14	14.00	1.99	
08-O-Oxygen	16	16.00	2.00	3.99
09-F-Florine	18	19.00	3.00	
10-Ne-Neon	20	20.18	1.18	4.18
11-Na-Sodium	22	22.99	2.81	
12-Mg-Magnesium	24	24.31	1.32	4.13
13-Al-Aluminum	26	26.98	2.67	
14-Si-Silicon	28	28.09	1.11	3.78
15-P-Phosphorus	30	30.97	2.88	
16-S-Sulfur	32	32.07	1.10	3.98

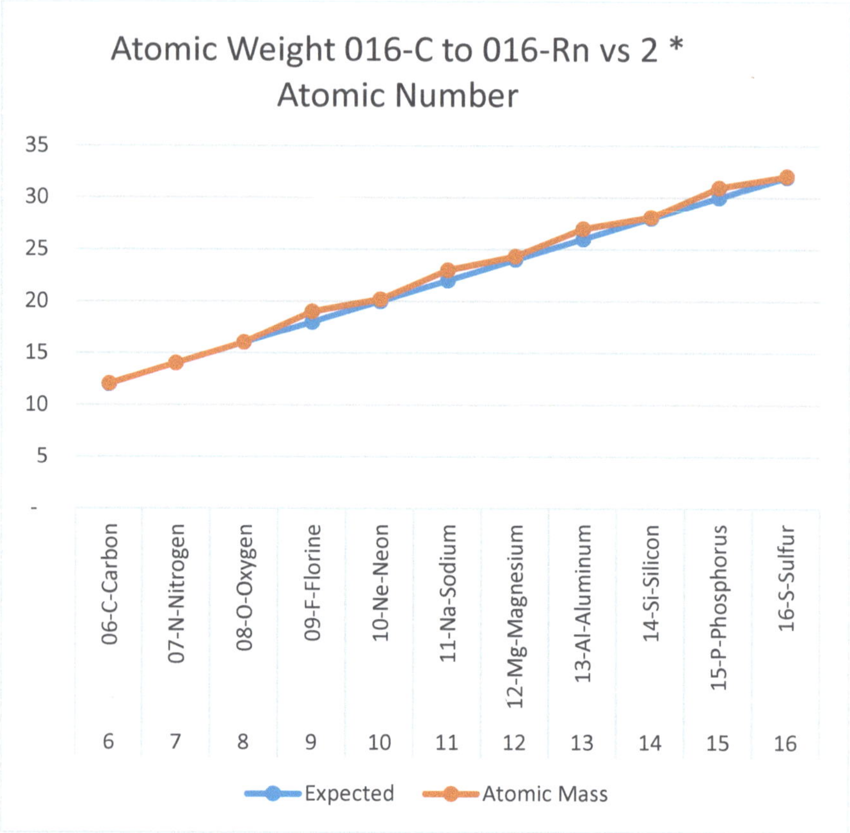

Atomic Weight 016-C to 016-Rn vs 2 * Atomic Number
35
30
25
20
15
10
5
-
06-C-Carbon
07-N-Nitrogen
08-O-Oxygen
09-F-Florine
10-Ne-Neon
11-Na-Sodium
12-Mg-Magnesium
13-Al-Aluminum
14-Si-Silicon
15-P-Phosphorus
16-S-Sulfur
6
7
8
9
10
11
12
13
14
15
16
Expected
Atomic Mass

The odd number of protons get an extra neutron to separate the extra one as the ring comes back to the beginning.

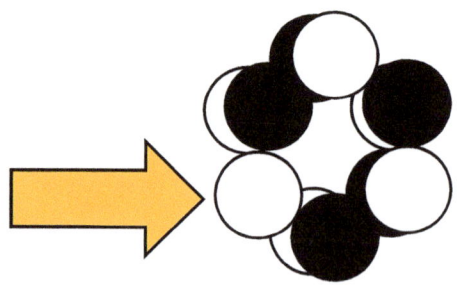

Without the extra neutron, one odd proton counts, the ring will end with proton meeting proton – and nuclear decay.

After that level of 2:1 016-Sulfur to 32 nuclear particles, the structure probably gets too wiggly, and one side could hit the other proton-to-proton, and viola, nuclear decay.

That next set of logic work very well up to 6 around x 7 rings as alternating if that were built perfectly.

However, what really is that two pairs of protons and neutrons need a fifth proton to assure that protons are separate.

That means that the structure then start building with the extra neutron every two protons into 5:2 or 2.5:1 ratio. A whole electron then binds into something like:

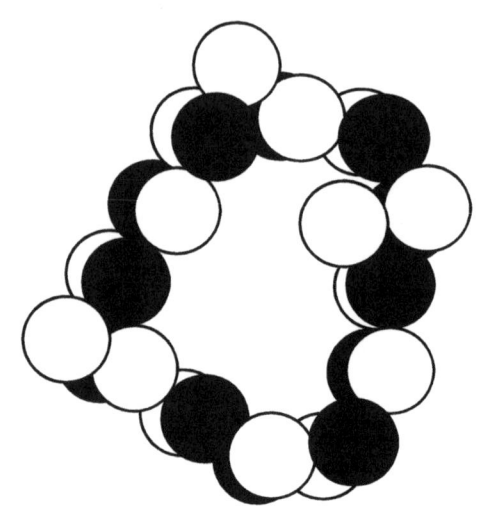

Symbol-Element Name	Expected	Atomic Mass	Change from Last	Sets of 2
11-Na-Sodium	22	22.99	2.81	
12-Mg-Magnesium	24	24.31	1.32	4.13
13-Al-Aluminum	26	26.98	2.67	
14-Si-Silicon	28	28.09	1.11	3.78
15-P-Phosphorus	30	30.97	2.88	
16-S-Sulfur	32	32.07	1.10	3.98
17-Cl-Chlorine	34	35.45	3.38	
18-Ar-Argon	36	39.95	4.50	7.88
19-K-Potassium	38	39.10	(0.85)	
20-Ca-Calcium	40	40.08	0.98	0.13
21-Sc-Scantium	42	44.96	4.88	
22-Ti-Titanium	44	47.88	2.92	7.80
23-V-Vanadium	46	50.94	3.06	
24-Cr-Chromium	48	52.00	1.06	4.12

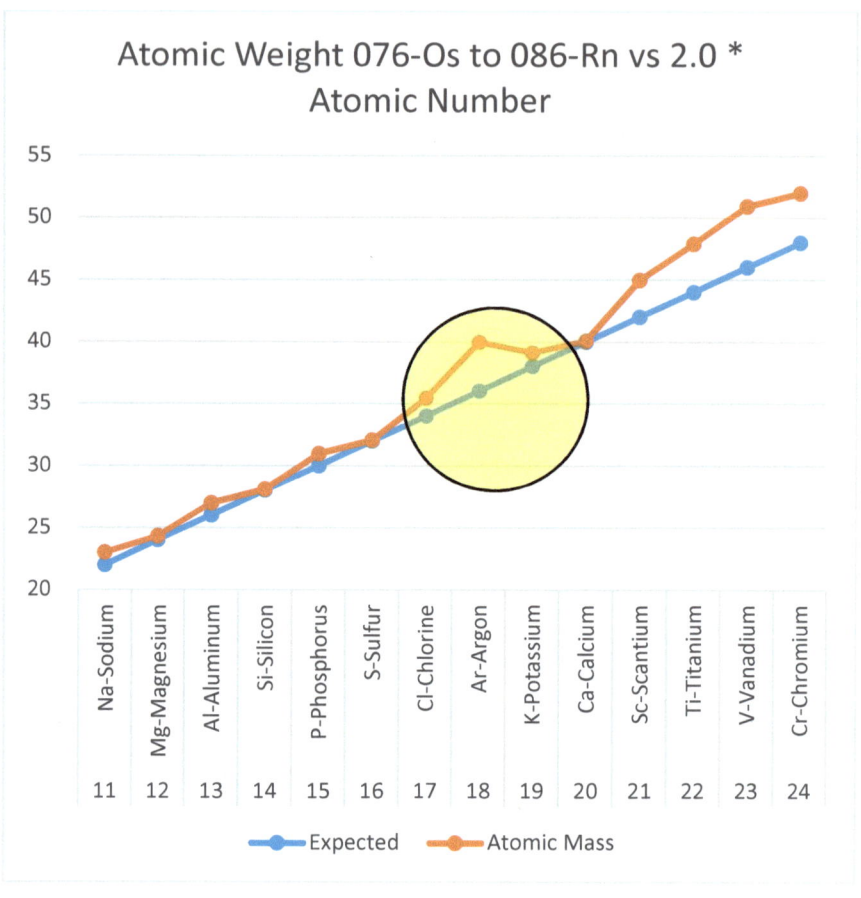

Atomic Weight 076-Os to 086-Rn vs 2.0 * Atomic Number

You can observe that for this nucleus physical structure, the Atomic Weight increases by three (3), then by one (1), then by three (3), and then by one (1) again. It is a clear pattern.

In my AVSC Scrunched Cube Atomic Model, this is 'three-then-one' building is a double ring. The neutron must get added first, and in both directions and rings. The double ring gets a proton-neutron pair, but that would make the ends meet proton-proton, so the third particle, the neutron separator must occur for odd numbered double-ring nucleus.

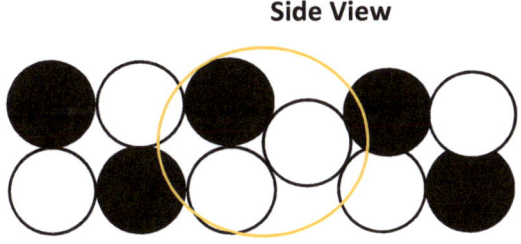

It must be a set of three.

Of course, to get from that to an even-count Atomic Weight, you only need a proton from the correct direction.

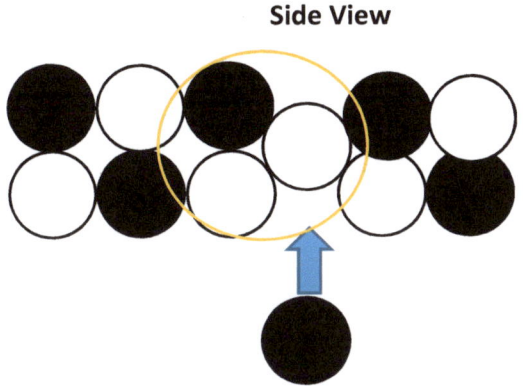

So, it becomes alternating proton-neutron, with a 2nd layer neutron-proton. It is very stable.

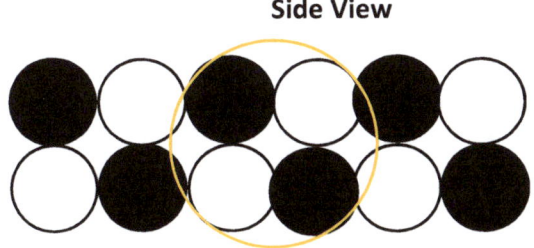

That is how from even to even becomes three (3), then one (1) with two neutrons needed to add two protons to a nucleus.

Yet something changes in the next section. Eventually, the single ring gets too big, and wobbly. Then, the double ring gets too big, and wobbly. Based upon this the number around a ring before it gets to big is:

Atomic Number	Symbol-Element Name	Expected	Atomic Mass	Change from Last	Sets of 2
11	Na-Sodium	22	22.99	2.81	
12	Mg-Magnesium	24	24.31	1.32	4.13
13	Al-Aluminum	26	26.98	2.67	
14	Si-Silicon	28	28.09	1.11	3.78
15	P-Phosphorus	30	30.97	2.88	
16	S-Sulfur	32	32.07	1.10	3.98
17	Cl-Chlorine	34	35.45	3.38	
18	Ar-Argon	36	39.95	4.50	7.88
19	K-Potassium	38	39.10	(0.85)	
20	Ca-Calcium	40	40.08	0.98	0.13
21	Sc-Scandium	42	44.96	4.88	
22	Ti-Titanium	44	47.88	2.92	7.80
23	V-Vanadium	46	50.94	3.06	
24	Cr-Chromium	48	52.00	1.06	4.12

The structure works 2:1 all the way to 16-S-Sulfur. Perfect to 32 particles. 2 * 2 * 2 * 2 * 2. Two rows of 16, or 4 rows of 8 around.

Yet, something happens at 18-Ar-Argon with 36 that makes the structure jump, then the next Elements cannot jump, but bind more protons. Yet is back to the perfect radio at 20-Ca-Calcium again.

Now, 36 is one of those magic numbers. It is 2 * 2 * 3 * 3. It could be the nice limit going from 2 * 18 = 36 particles. Or it could be 3 * 12 = 36. Or, it could be 4 * 9 = 36 particles.

Something happens at 21-Sc-Scantium that makes the next section use up eight (8) particles to add two protons.

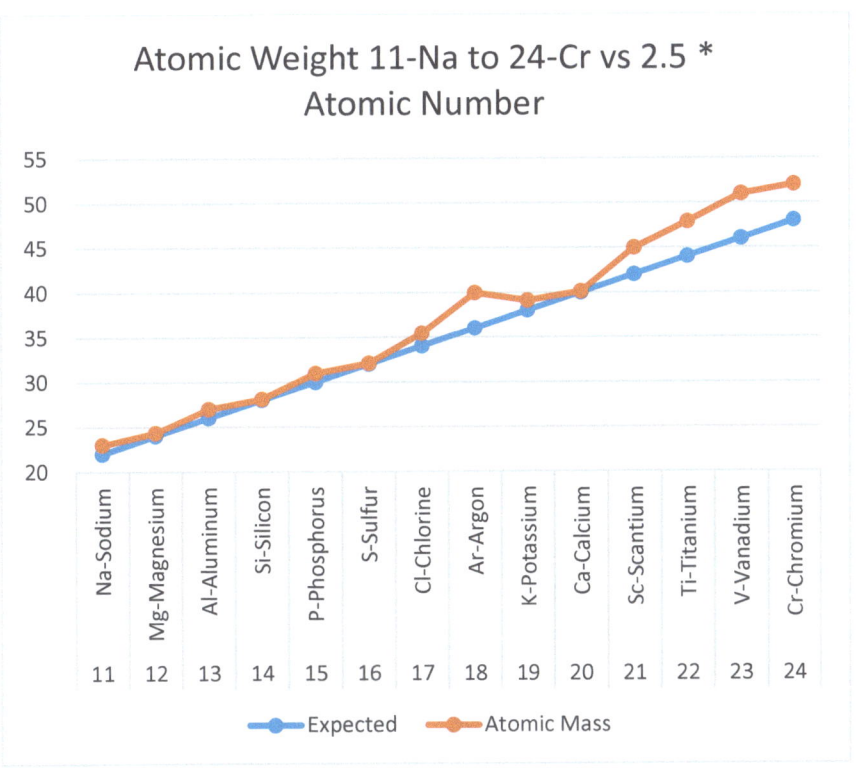

There is a strange bump, then a permanent separation where the 2.0:1 becomes 2n + 4 for even-proton-count Elements and 2n+5 for odd-proton-count Elements. Therefore, every other layer built needs two neutrons to separate the ring layer. That special connector section is not 2:0, but 6 neutrons to connect 2 or 3 protons.

To get to a wider cylinder of nucleus particles, the structure cannot build as easily. Every so often, an extra neutron set of spacers occurs.

If we have:

                    N   P   N

Meeting

                    N   P   N

The only want that works is by:

                    N   P   N

                        N   N

                    N   P   N

There is six-neutrons to connect two protons. Four extra neutrons are necessary to handle the end to end match.

Side View

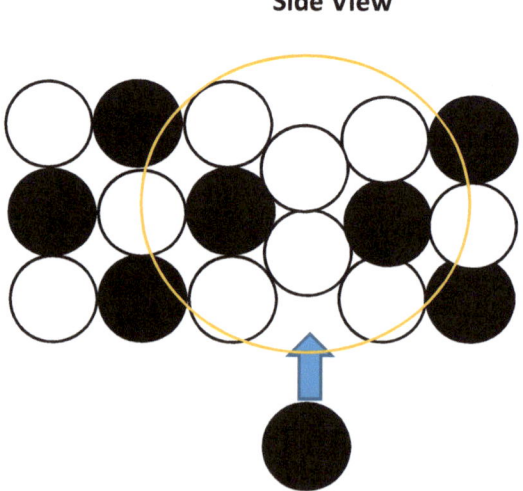

However, when that gets another proton, it can only be from one side.

Side View

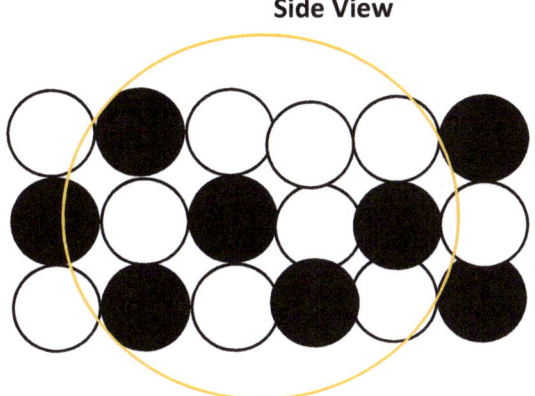

You can see this as a cylinder of 6 x 6 which is 36, but caps at 6 x 7 = 42. These logic work very well up to 6 around x 8 rings as alternating if that were a perfect.

NOT POSSIBLE

In the larger configuration, the rings build on the outside, so they are much large. An atom needs extra neutrons to maintain separation. This structure of building needs three neutrons to make a stable base for the proton to build upon.

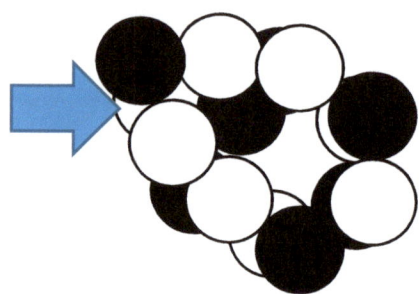

Again, I suggest that every student tries this with a set of magnetic balls. You can learn what can work and cannot work given the force of magnetics. Remember:

Nucleomagnetics is the most powerful force inside the Bohr radius.

A similar process can occur in another building direction; the cylinder can get thicker.

Side View

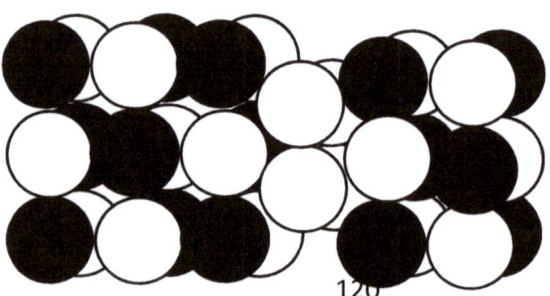

Around the ring, a nucleus then has a cylinder, empty in the middle.

3 wide x 2 deep = 6

3 wide x 3 deep = 9

This intermediate structure creates the method to get from 36 to 72 particles.

This leads to a section of where the Atomic Weight (protons plus neutrons) increase versus the Atomic Number (protons) at a new rate of 2.5:1.

Symbol-Element Name	Expected	Atomic Mass	Change from Last	Sets of 2
24-Cr-Chromium	53	52.00		
25-Mn-Manganese	55	54.94	2.94	
26-Fe-Iron	58	55.85	0.91	3.85
27-Co-Cobalt	60	58.93	3.09	
28-Nickel	63	58.69	(0.24)	2.85
29-Cu-Copper	65	63.55	4.85	
30-Zn-Zinc	68	65.38	1.83	6.69
31-Ga-Gallium	70	69.72	4.34	
32-Ge-Germanium	73	72.63	2.91	7.25
33-As-Arsenic	75	74.92	2.29	
34-Se-Selenium	78	78.97	4.05	6.34
35-Br-Bromine	80	79.90	7.27	
36-Kr-Krypton	83	83.80	3.90	11.17
37-Rb-Rubidium	85	85.47	6.50	
38-Sr-Strontium	88	87.62	2.15	8.65

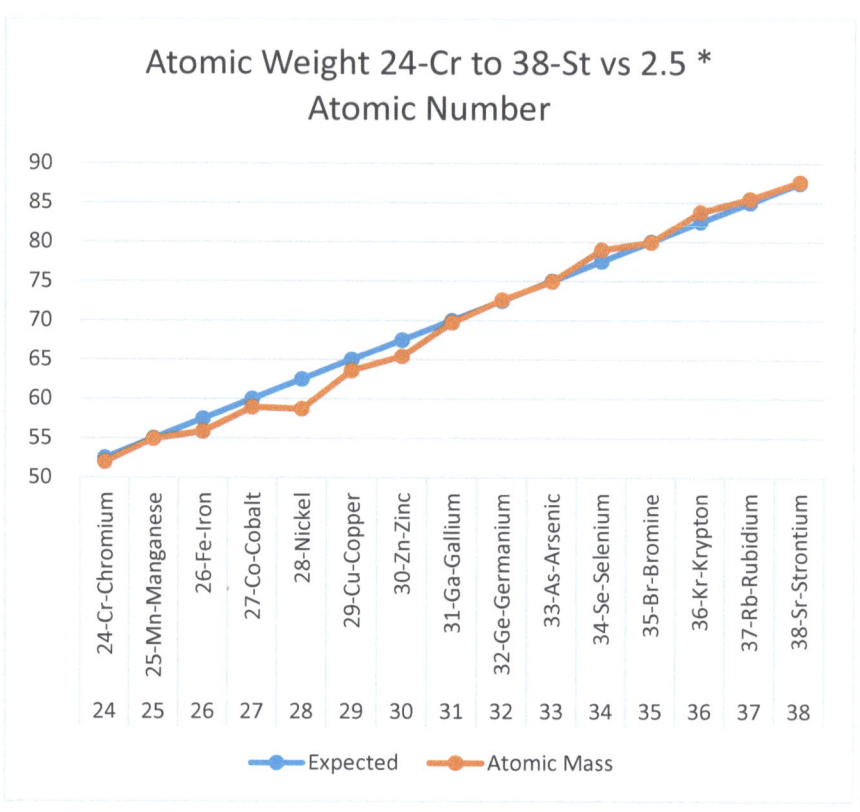

Yet, some of the original structure also needs extra neutrons, to fit into the 3D cylinder. Therefore, extra neutrons are needed until the entire structure is at particles:protons of 2.5:1.

This continues to a range where the additional are clearly 5 particles for every 2 protons (2.5:1):

Symbol-Element Name	Expected	Atomic Mass	Change from Last	Sets of 2
38-Sr-Strontium	88	87.62		
39-Y-Yttrium	91	88.91	1.29	
40-Zr-Zirconium	93	91.22	2.32	3.60
41-Nb-Niobium	96	92.91	1.68	
42-Mo-Molybdenum	98	95.95	3.04	4.73
43-Tc-Technetium *	101	**98.00**	2.05	
R	103	101.07	3.07	5.12
45-Rh-Rhodium	106	102.91	1.84	
46-Pd-Palladium	108	106.42	3.52	5.35
47-Ag-Silver	111	107.87	1.45	
48-Cd-Cadmium	113	112.41	4.55	5.99

One interesting note is that the section at the end of the low increase is radioactive. The postulate that a full layer (width or depth of rings) gets to a limit. It tries to keep that smaller structure, but that leaves a stray protons exterior at a new layer alone. As a result, it is easily dislodged, and radioactive decay back to the stable.

This happened at 18-Ar-Argon with a strange bump at the end of the perfect 2.0:1 range. With 36 particles, it is a great range of 2*2*3*3 which means the structure can be 4 x 9, 6 x 6 (my guess), or 2 x 18.

Here the radioactive, linked to a position that change from at low ratio section (even-to-even < 4) moving to a clear (even-to-even >5 and even 6). The Element at that transition has trouble staying together. We will re-visit radioactivity much deeper. For large atoms, it really gets really interesting.

The interesting observation about this transition point is that it occurs at 100 particles, which is 2*2*5*5, which could be 4*5*5 a well-balanced limit. Or it could be 108 which is 2*2*3*3*3 less the open core section of 9 for 99 particles as the structure limit.

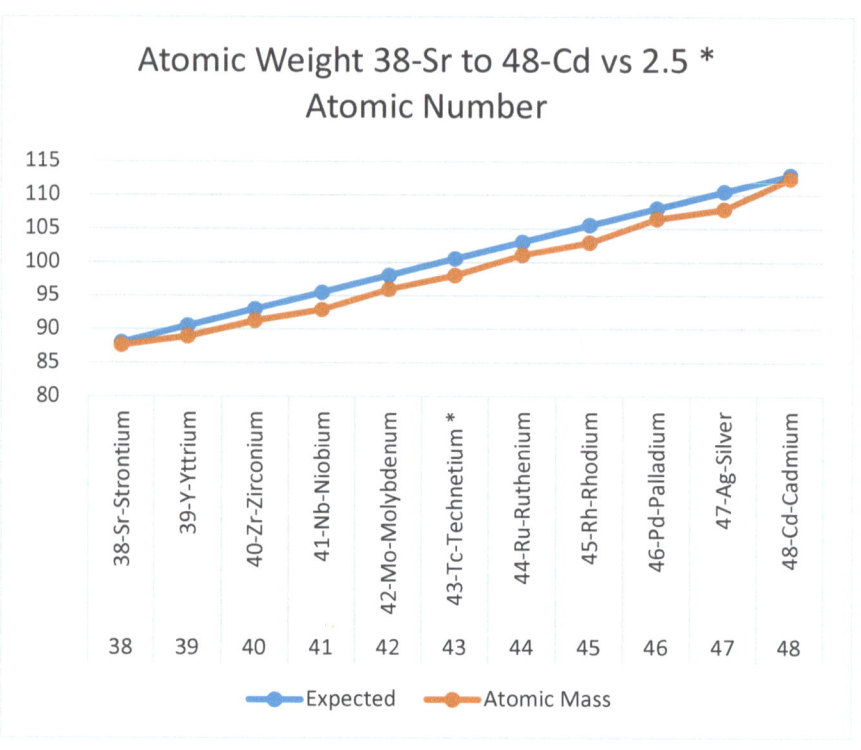

Atomic Weight 38-Sr to 48-Cd vs 2.5 * Atomic Number

* 43-Tc-Technetium is radioactive.

The growth continues at 2.5:1 which includes another radioactive point at 144 particles. Notice that 144 = 2*2*3*3. Alternatively, you can this of that as 2*3*5*5 = 150 less that core of 5 particles for 145.

In either case, there is a sudden slow in the particle growth ending with a radioactive structure, then the section after returns to a 2.5:1 growth rate from the base particle count (with those previous structure as the base).

* 61-Pm-Promethium is radioactive.

Symbol-Element Name	Expected	Atomic Mass	Change from Last	Sets of 2
49-In-Indium	117	114.82		
50-Sn-Tin	119	118.71	3.89	
51-Sb-Antimony	122	121.76	3.05	6.94
52-Te-Tellurium	124	127.60	5.84	
53-I-Iodine	127	126.90	(0.70)	5.14
54-Xe-Xenon	129	131.29	4.39	
55-Cs-Cesium	132	132.91	1.61	6.00
56-Ba-Barium	134	137.33	4.42	
57-La-Lanthanum	137	138.91	1.58	6.00
58-Ce-Cerium	139	140.12	1.21	
59-Pr-Praseodymium	142	140.91	0.79	2.00
60-Nd-Neodymium	144	144.24	5.34	
61-Pm-Promethium *	147	**145.00**	0.76	6.10
62-Sm-Samarium	149	150.36	9.45	
63-Eu-Europium	152	151.96	1.60	11.06

In the below, you can see the bulge above the line, then flatting and the radioactive element as the turning point. Further, the radioactive elements get found at point below a certain ratio.

It is especially pronounced near these 2 and 3 combinations. In quantum mechanics, these are fancy step functions where 'color' determines when it is functionality changes at a 2, and when a 3 works.

In AVSC, it is simply the famous map challenge, to draw a map of protons where the same type cannot touch. The great challenge is building the no-touch protons is a 3D map.

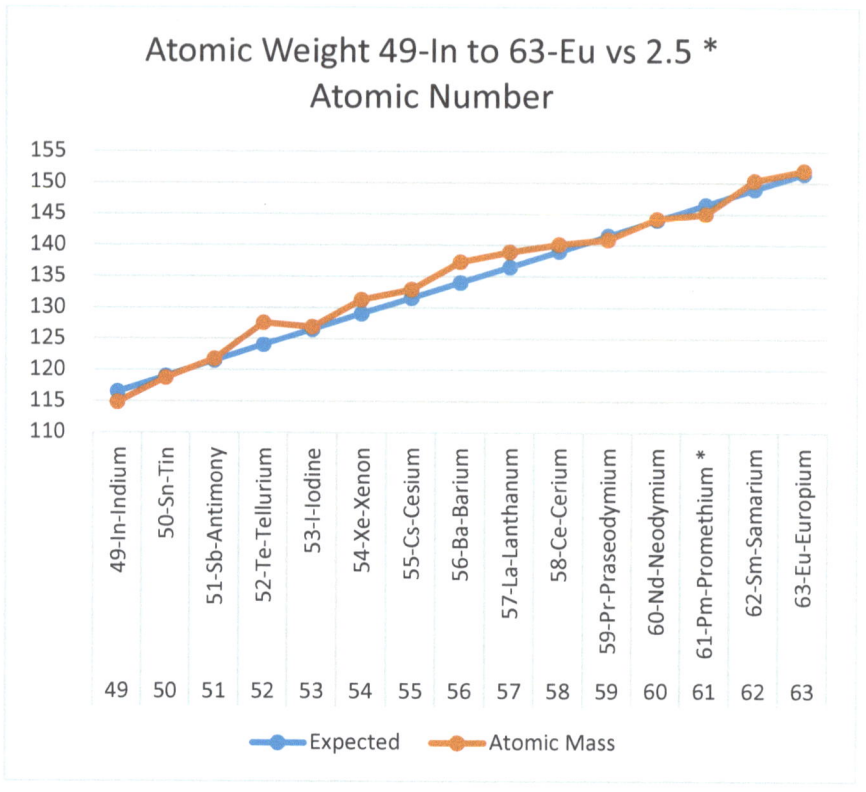

Atomic Weight 49-In to 63-Eu vs 2.5 * Atomic Number

This is a stable process for this range of elements. However, the change place is at 61-Promethium. Notices that 61- is not where electrons shells change

In AVSC, electron shells get full and start new shells in spherical 3D hemispheres.

In AVSC, nucleus layers change in cylinder width and depth with an open core.

The full layers in nucleus structure has no relation to the full shells of electrons.

Finally, there begins the section where the number of neutrons required to bind a proton without touching another proton becomes even greater (11:1).

In AVSC, this is the depth/thickness of the cylinder becoming a third layer. That is the structure becomes 3 layers, open center, 3 layers. That will become important as the final stable fully-loaded structure of a nucleus gets created at 216 less 6 for 210 particles.

Symbol-Element Name	Expected	Atomic Mass	Change from Last	Sets of 2
64-Gd-Gadolinium	157	157.25		
65-Tb-Terbium	160	158.93	1.68	
66-Dy-Dysprosium	162	162.50	3.57	5.25
67-Ho-Holmium	165	164.93	2.43	
68-Er-Erbium	167	167.26	2.33	4.76
69-Tm-Thulium	170	168.93	1.68	
70-Yb-Ytterbium	172	173.05	4.12	5.80
71-Lu-Lutetium	175	174.97	1.91	
72-Hf-Hafnium	177	178.49	3.52	5.44
73-Ta-Tantalum	180	180.95	2.46	
74-W-Tungsten	182	183.84	2.89	5.35
75-Re-Rhenium	185	186.21	7.72	
76-Os-Osmium	187	190.23	4.02	11.74
77-Ir-Iridium	190	192.22	8.38	
78-Pt-Platinum	192	195.08	2.87	11.24

You can see the graph of this range begin to race upwards after 74-W-Tunsten with 184 particles. This number being 192 (2*2*2*2*2*3) less 8 for the core.

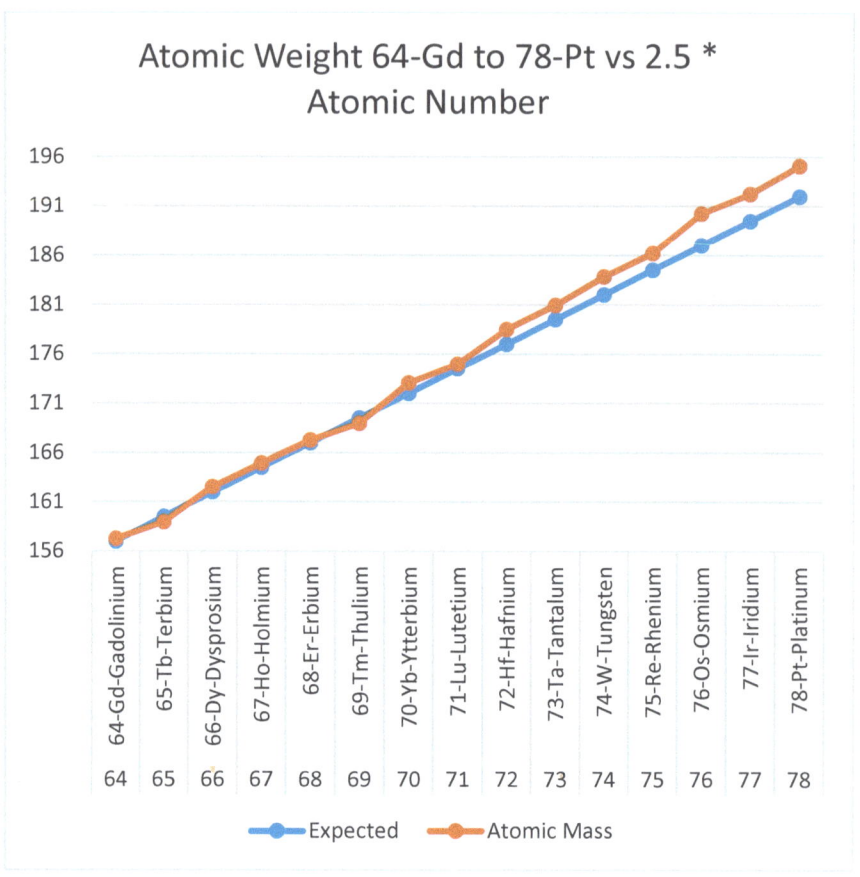

Central Channel of Nucleus Remains Empty along its induced nucleomagnetics axis because avoiding protons as the center is almost impossible.

A position along the nucleomagnetics axis touches two positions along the nucleomagnetics axis, and another six position around the cylinder. A proton cannot avoid eight positions that might have protons. The natural solution becomes to build structures where the axis row is empty.

What are the Structures of Protons and Neutrons as Build the Nucleus of Larger Molecules?

Because of the nature of these proton building, we see different types of jumps between the most stable isotopes of each Elements. These changes are based upon the size and shape of the base nucleus proton-neutron configuration.

Ratio: 2.0:1
Group Jumping by four (4) particles – two (2) protons plus two (2) neutrons creates stable growth:

Group Jumping by four (5) particles – two (2) protons plus two (2) neutrons creates stable growth:
Ratio: 2.5:1

5)   Even to odd – jumping by three (3) for the extra neutron at the end bridging two layers

6)   Odd to even – jumping by one (1) making the two

Group Jumping by more than five (5) particles – two (2) protons plus two (2) neutrons creates stable growth: *Ratio: 2.5:1*

7) Even to odd – jumping by three (3) for the extra neutron at the end bridging two layers

8) Odd to even – jumping by one (1) making the two

Finally, the last stable group and the start of the full radioactive heavy elements.

Symbol-Element Name	Expected	Atomic Mass	Change from Last	Sets of 2
76-Os-Osmium	190	190.23		
77-Ir-Iridum	193	192.22	1.99	
78-Pt-Platinum	195	195.08	2.86	4.85
79-Au-Gold	198	196.97	1.89	
80-Hg-Mercury	200	200.59	3.62	5.51
81-Tl-Thallium	203	204.59	4.00	
82-Pb-Lead	205	207.20	2.61	6.61
83-Bi-Bismuth	208	208.98	1.78	
84-Po-Polonium	210	**209.00**	0.02	1.80
85-At-Astatine	213	**210.00**	1.00	
86-Rn-Radon	215	**222.00**	12.00	13.00
87-Fr-Francium	218	**223.00**	1.00	
88-Ra-Radium	220	**226.00**	3.00	4.00

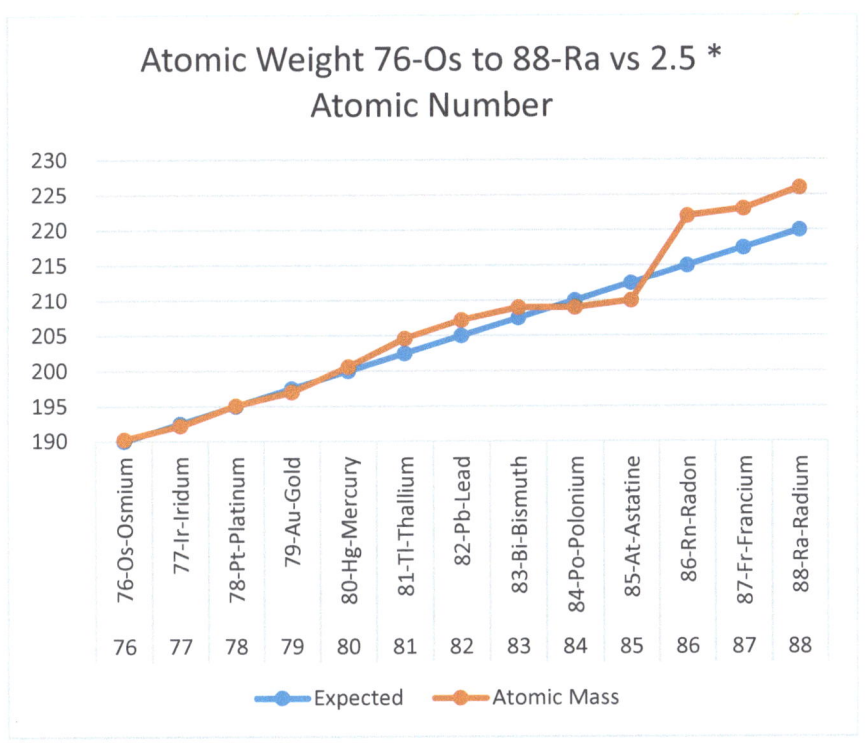

Atomic Weight 76-Os to 88-Ra vs 2.5 * Atomic Number

There it is that budge, then dip below the 2.5:0 ratio, and that is the radioactivity transition point.

In AVSC, the 3D particle layered structure is 216 less 6 = 210, and the 216 is 2*2*2*3*3*3 or 6*6*6.

Why does a Nucleus Become Unstable and Decay?

There is point after 082-Lead which has 210 Atomic Weight, that the nucleus of any atom of a higher Element are unstable, and decay with a mathematical certainty of rate.

Something magically happens at the Atomic Weight of 210. You can see something changes in the structure.

Symbol-Element Name	Expected	Atomic Mass	Change from Last	Sets of 2
83-Bi-Bismuth	208	208.98	1.78	
84-Po-Polonium	210	209.00	0.02	1.80
85-At-Astatine	213	210.00	1.00	
86-Rn-Radon	215	222.00	12.00	13.00
87-Fr-Francium	218	223.00	1.00	
88-Ra-Radium	220	226.00	3.00	4.00
89-Ac-Actinium	223	227.00	1.00	
90-Th-Thorium	225	231.04	4.04	5.04
91-Pa-Protactinium	228	231.04	-	
92-U-Uranium	230	238.03	6.99	6.99

Also, notice that after the transition point the next structure is 4 particles for 2 protons. It is a hanging chain just like the very small chain and single ring nucleus structures.

Of course, after that it gets more difficult to find a spot quickly with 5:1, then 6:1 ratios.

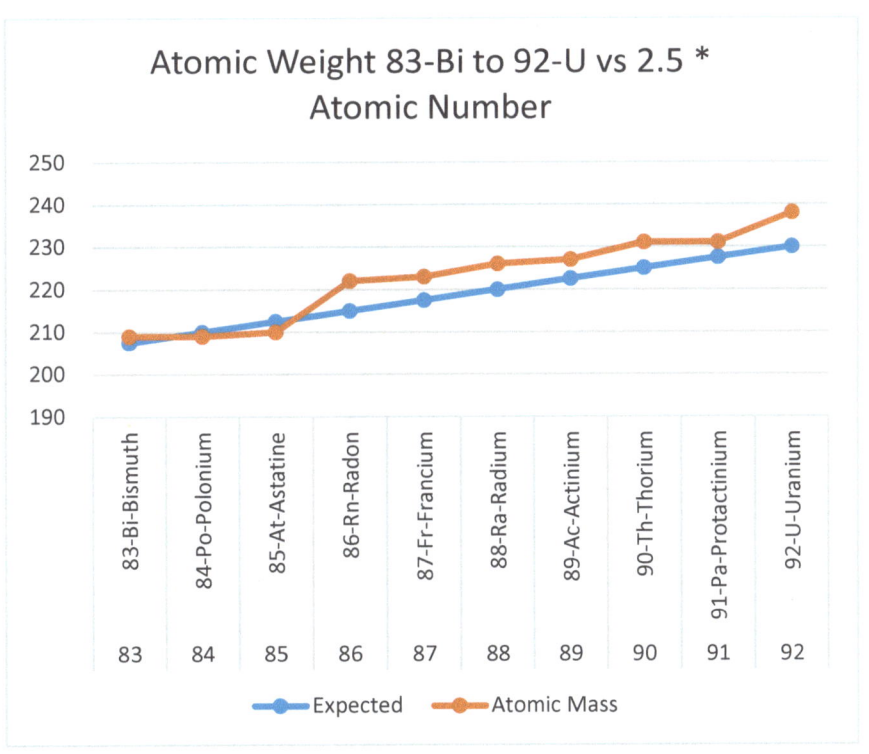

Atomic Weight 83-Bi to 92-U vs 2.5 * Atomic Number

At 086-Rn-Radon, elements become radioactive. Radioactive elements tend to decay back to an elements like 086-Pb-Lead. There is something fundamental happening at that number of protons, the Atomic Number, and Atomic Weight, the number of nucleus particles, the sum of protons and neutrons.

Nucleus Decay

We know that most nucleus remain stable once in the proton-neutron-proton chain-ring configuration. Yet, we also know that very large molecule do change to smaller molecules at a very consistent percentage rate. We can measure the age of fossils by calculating the nucleus decay of Carbon-14 to determine age, even over centuries of time.

In nucleus decay, a percentage of the oversized atoms decays. However, as that happens that percentage starts lower for the next period. That means there is less to atoms in round two. This goes on and on. As a result, the amount of material

The basic idea is that external forces, like particles, waves, or changes in heat, move the nucleus particles in their structure. If that movement knocks the separating neutron out of place, or delivers a proton directly to a proton, then both repel, taking a few surrounding particles out of the nucleus. This is called nucleus decay.

An additional postulate says that if the basic components of a neutron break down, then that leaves three protons touching each other, and of course nucleus decay. We will not focus on neutron decay in this volume; that is left to others.

Radioactivity

We know that most nucleus remain stable once in the proton-neutron-proton chain-ring configuration. Yet, we also know that very large molecule do change to smaller molecules at a very consistent percentage rate. We can measure the age of fossils by calculating the nucleus decay of Carbon-14 to determine age, even over centuries of time.

In nucleus decay, a percentage of the oversized atoms decays. However, as that happens that percentage starts lower for the next period. That means there is less to atoms in round two. This goes on and on. As a result, the amount of radioactive material decrease exponentially. This means that as some base material (B) decays at a rate (r), the amount of radioactive material (M) will be decreasing, but only based upon the amount left, so it has that slow slope to zero.

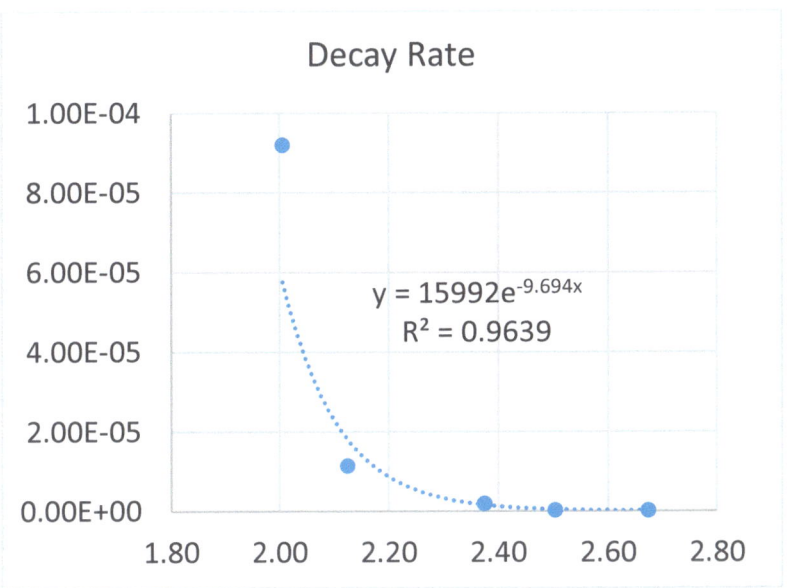

The above graph is important, because it is a primary evidence of the AVSC nucleus model. That graph is a scatter diagram comparing a) various radioactive elements and isotopes versus b) the ratio of particles to protons in the outermost, exposed layer.

Element	Neutron Count	Excess	Ratio	Decay Rate
89-Th-Thorium	142.04	16.04	2.67	1.41E-07
91-Pa-Protactinium	140.04	14.04	2.01	9.20E-05
92-U-Uranium - average	146.03	20.03	2.50	2.49E-07
U-Uranium- Isotope-235	143.00	17.00	2.13	1.14E-05
U-Uranium- Isotope-237	145.00	19.00	2.38	1.87E-06

The Arno Vigen Scrunched Cube (AVSC) nucleus model is physical and geometric. It takes the 'road less traveled' versus the statistical, quantum path, and in that way can solve certain challenges better. Of course, quantum mechanics works for its set of functions, statistics, and harmonics are great. In the AVSC Nucleus Model for radioactive Elements:

- All the connections have at least one neutron physically in 3D separating each proton. Neutrons are initial 1:1 making the structure 2:1 total particles to protons. However, as the layers of rings increases, additional neutrons are required to keep separation. The ratios grows to 3:1, even 6:1 in steps.

- The inner layers are tightly formed, and thereby not statistically likely to get dislodged as radioactivity. The full inner layers are structures of 2, 3, 4, 5, or 6 wide and deep, probably in cylinders, with the most inner 'row' empty. It is shaped like a cylinder with extra chains hanging outermost.

- The nucleomagnetics of all the particles do not combine exactly to the magnetics of the full structure. First, because these are magnetics, the combination of 1-ring creates a perpendicular nucleomagnetics axis. A continuous ring of magnetics generates a force to the side even if half of the magnetics go one direction, and half go the other. Second, that strength is less than the sum of the particle's fields because in 3D some cancel each other. Further, the math of 2-ring with natural opposite direction creates a different calculation of the perpendicular.

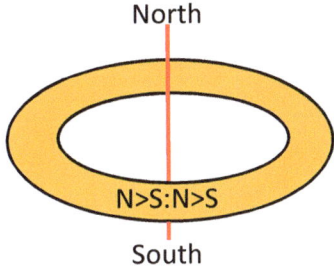

North

N>S:N>S

South

- The most outer layers has difficulty finding a stable connection locations. As a result, the most outer layer can become either these chains which flop around (and get near other protons to decay).

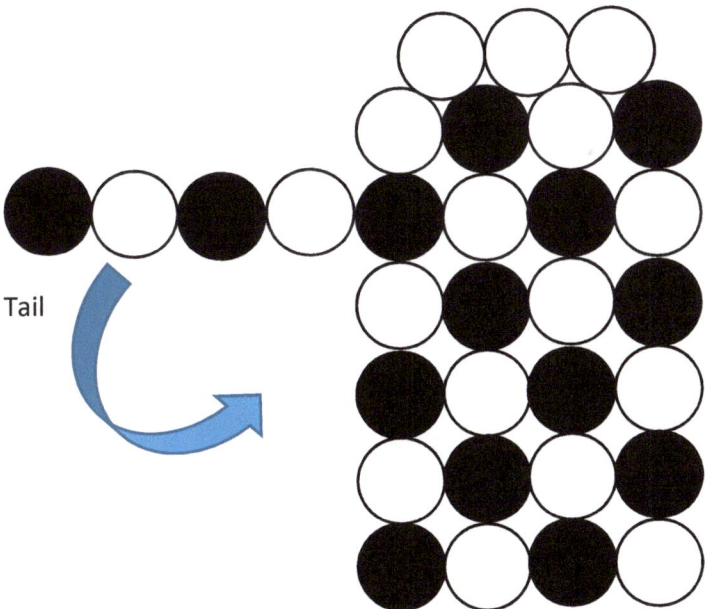

Tail

Or these extras are weakly linked partial layers

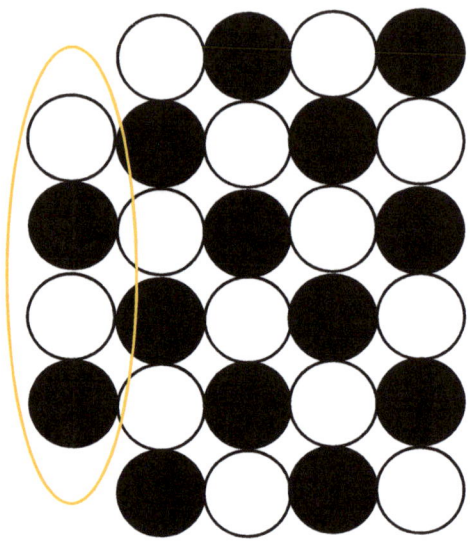

Which might shift enough to create proton-proton repulsion for decay.

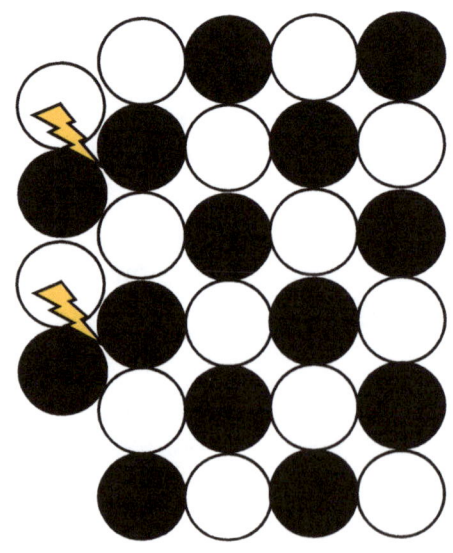

- This leads to a relationship of the outer ratio of total particles versus the protons that determines the radioactive decay rate in the known time exponential formula. If the extra layer is 2:1 (equal neutrons versus protons), the minimum, the decay rates is huge. As the ratio gets to 2.5:1, the decay is back toward a stable nucleus, and takes years instead of minutes.

- Atoms along would not repeat decay. It takes the ejected particles from one radioactive atom to inject, and create decay in the next.

- This outer layer particles:protons ratio is only the minor factor. The important factors to calculate decay rates are:
    - The structure and forces of the approaching particle or particle-group
    - The speed, rotation, and orientation of both targets atoms and approaching particles
    - AVSC Nucleomagnetics Angle of particle-group delivery relative to the axis and the protective electron subshells
    - The density, and number of the atoms that are radioactive in the target space.
    - The ratio of nucleus outer-layer particles

- The breaks in major layers of the nucleus are not at the same places (as breaks in the electrons shells (2, 8, 8, 18, 18, 32, and 32).

    - The nucleus builds with physical magnetic connections without hemispheres, in rings and layers.
    - The electron shells build without connections, electrons repel each other, in hemispheres based upon a central pull (nucleus) and nucleomagnetics angles for subshells.

Conclusions

By finding the nucleus structure of each element, isotope, and isoform. You have a clean basis that can concretely determine:

- Nuclear decay rates, including the energy needed to start a nuclear reaction

- Calculate nucleomagnetics moments, especially in those forms where nucleomagnetics fields flow in opposite direction

However, there are many more questions to study. Please become an explorer with me.

Hugs for everyone.

E Arno Vigen

Glossary

Atom – The balanced combination of a nucleus consisting of P protons and N neutrons, with a shell of P electrons.

Bohr radius – The ratio at which the nucleomagnetics field and the charge-force field balance based upon a single-proton 001-H Hydrogen atom.

$$5.2917721092171717 \times 10{-11}m$$

Coulomb – the measurement of the charge-force of proton or electron particles.

Coulomb's Law

$$F = k_e \frac{q_1 q_2}{d^2}$$

Electron – an atomic particle with a negative (-) charge with a repulsion to nucleomagnetics fields.

Isotope – a class of atoms with structure of a nucleus which have the same number of protons, but not the same number of neutrons.

Isoform – a class of atoms with structure of a nucleus which have the same number of protons AND the same number of neutrons.

Nucleomagnetics field – a field around each base particle, protons and neutrons, which has an axis and where the field strength changes/increases based upon a function of the inclination/longitude angle to that axis. Further, the strength is the same for every longitude/azimuth angle. Further, the nucleomagnetics has the same sign in both direction (versus different in Motomagnetics and thereby North-South).

Motomagnetics field – the north-south oriented field associated with a proton or neutron. The grouping nucleomagnetics field may have different orientation, strength than the nucleomagnetics field of the individual particles.

Nucleomagnetics ring - A structure where magnets link north-to-south in a chain then connected at the distant end. As such, that field actually creates a further nucleomagnetics field perpendicular to the ring.

Neutrino – an atomic particle without charge and possibly without magnetism.

Neutron - an atomic particle without charge and with magnetism. Currently research indicates it may be a combination of subparticles.

Nucleus particles – particles that bind by nucleomagnetics in the nucleus. These are positive-charged (+) protons and no-charge neutrons.

Proton - an atomic particle with a positive (+) charge and with magnetism.

Arno Vigen Scrunched Cube Atomic Model Postulates:

For more than a century, the pendulum of physical sciences, especially the study of the nucleus, moved away from Newton and the concrete, physical world. The Arno Vigen Scrunched Cube Atomic Model postulates below move the pendulum one step back towards center. A is A. Physical reasons are better describers of physical science, and when those physical factors are discovered, understanding is easy.

While other solutions get the correct answer in brilliant, amazing, creative formulas, the deep answer becomes simple and real. It becomes something that we can teach to every student, without LaGrange, without Hamiltonians, without Schrodinger imaginary numbers, and with Gaussian differentials -- without, or better said resolving into the basics of electrostatics charge force and a base magnetic force, instead of the invented 7, 9, 11 or 18 dimensions of the latest version of string-theory.

The AVSC Atomic Model postulates go back to the basics:

- Three physical dimensions (length, width, height or their spherical equivalents)

- Time

- Electrostatic and base magnetic fields and force*

- Known particles – protons, electrons, and neutrons

These, electric-charge-force and magnetism are understood as intimately linked, but the full details of that linkage will wait until someone discovers the direct interconnection in physical dimensions.

Each of the postulates takes current, complex calculations, and gives them a clean, deterministic, Newtonian path using only the above basics.

#1 - Every nucleus holds together via a chain/ring-magnet organized as proton-neutron-proton-and so on:

- Resolves what holds the nucleus together. This 'strong force' gets based upon when alternative particles, in a chain, connect along an oriented nucleomagnetics field, so that, at the nucleus distance, magnetism is stronger than charge-force if the positive (+) charged protons are separated enough by an intermediate neutron.
- Educational nucleus-plus-chemistry set US patent 15/256865 pending relating the nucleomagnetics field of the particles to the overall nucleomagnetics field of the chain or ring nucleus structure.

Figure 12

While most people think that Electrostatic Charge is stronger than magnetism, yet a nucleus stays together, the trick is that magnetics stay strong (do not decrease) when in a chain:

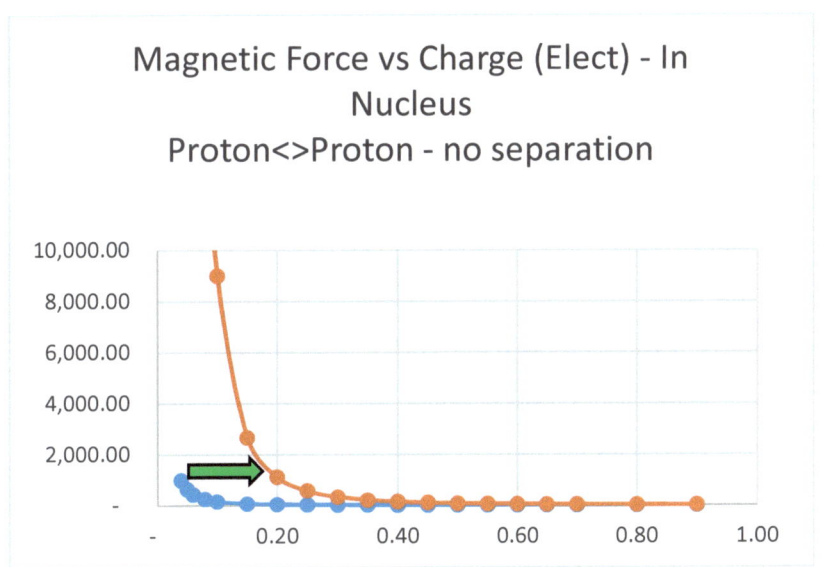

Because a nucleomagnetics field does not start decreasing until the physical chain is broken, there is proton-neutron-proton configurations that allow a nucleus to stay bonded together.

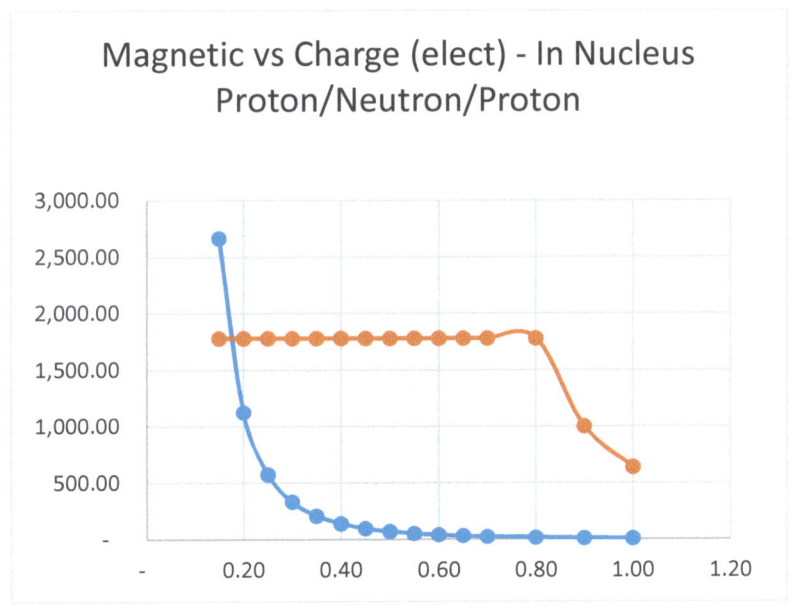

It simplifies to the above two graphs that explain the charge force versus the nucleomagnetics force. Both decrease, but for a particular point, the charge force is always stronger (chart 1), but in chains the charge is not a point, but a chain which means it does not decrease until the end of chain – keeping it strong enough to stay linked even if protons repel.

#2 Electrons repel nucleomagnetics – *at both poles*

- Resolves what force makes the electrons stay in a shell
- Resolves spin number in various subatomic particles
- Explains basic electricity
- Explains EMP pulses

#3 Atoms and molecules rotate and move as a group. Electrons positions in a shell is the balance of electrostatic attraction (1/distance-squared) versus nucleomagnetics repulsion (an inclination angle function x 1/distance-cubed). With the repulsion of nearby electrons, that makes the calculation of forces build in two hemispheres, and similar inclination angles (electron subshells) from the nucleomagnetics pole toward its equator.

- Replaces Bohr-Sommerfeld model and quantum Standard Model with geometric, deterministic view
- Replaces most applications of angular momentum in quantum mechanics; leaving only angular moment relative to complete sets as rotating.

#4 Electrons shells build in geometric forms in two hemispheres within the nucleus nucleomagnetics field with a) doubling because hemispheres, the circles (squared). The subshells are axis to equator, with exceptions, and not in the direct filling order of aufbau/Pauli:

- Makes understanding of Pauli exclusion as a simple geometric picture – an electron at opposite positions in two hemispheres

- Educational chemistry set US patent 15/245,326 pending
- Chemical production US patent 15/490,870 pending
- Resolves 006-C Carbon 109.5 angle versus 007-N Nitrogen 107.5 angle versus 008-O Oxygen 104.5 angle
- Resolves 027-Co Cobalt melting point
- Resolves 005-B Boron bonding angles at 120 degrees
- Resolves why only 1 4s electron for 029-Cu Copper and other transition metals
- Resolves electromagnetic spectrum evidence at 8, not 10 in d-shells
- Resolves various d-subshell anomalies of the Pauli-aufbau current postulate
- Replaces s/p/d/f with geometric m/e/c/t/v with e as intermediate in some elements. m2 = nucleomagnetics poles scrunched, e = equatorial, c6 = rest of cube with m2, t6 = tetrahedral blocks towards the nucleomagnetics poles, and such

North

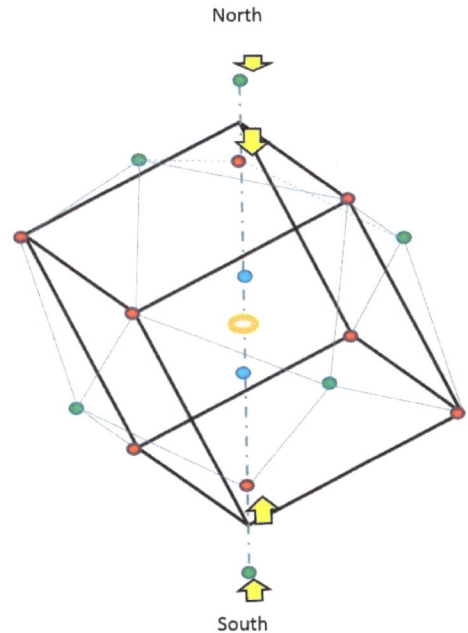

South

#5 Quantum mechanics is the harmonics of those particles pushing each other around relative to their natural (zero Kelvin) "settling" positions. All the quantum numbers can get determined from the AVSC model.

#6 Chemical Bonding resolved to geometric model.

- Electrons sharing relegated to special circumstances.
- Bond angles, bonding strengths determined
- Hydrogen bonding explained
- Contributing versus Receiving angles, radius distance, and positions differentiated for the same Element
- AVSC determines VSEPRS

#7 Radioactivity Explained. From the structure of a nucleus (#1 above), the root structure in 3D explains the decay rate of each Element and isotope.

#8 Gravity is the nucleus (proton, neutron) nucleomagnetics field reflected via an almost* universal electron shell radius (R_{ES}) as the electrostatic force net a) starting at the nucleus, versus b) the force for electrons starting just a little closer to distance objects over time.

- Permanently links gravity to the basic electrostatic and nucleomagnetics forces as a continuous function
- Resolves concrete link of nucleus 'atomic mass' particles (protons and neutrons) to 1/distance-squared observations of gravity
- Resolves mass loss in bonding by the bonds masking the nucleomagnetics portion electrons doing the bonding. Because they are inside the volume of the two nuclei that electron cannot contribute to R_{ES}, and as a result, the observed mass goes down.

- Replaces a number of the time-space warping calculations with physical factors that change R_{ES}, a physical distance so the calculation is knowable with physical dimensions

$$S = \frac{-M_A{}^2(z,n)_1(z,n)_2}{d^3(\frac{4}{3}\pi R_{ES}{}^3 + T)} = \frac{-(M_A(z,n)_1)(M_A(z,n)_2)}{d^3(\frac{4}{3}\pi R_{ES}{}^3 + T)}$$

$$mass = \sqrt{G}\ m = \frac{(\sqrt{G}\,)M(z,n)_1}{\frac{4}{3}\pi(R_{es})^3 + T} = \frac{M(z,n)_1}{\frac{4}{3}\pi(R_{es})^3 + T}$$

'Universal' will still have various Einstein, Lorentz adjustments as discussed, integrated, and potentially improved in other books now understood because based upon physical distance (R_{es}) which stretches, instead of space-time, as you get to a percentage of speed of light, and other factors.

The Continuous Fundamental Forces

1) Electrostatic Force

2) Nucleomagnetics Force

3) Motomagnetics Force –field movement disruptions create differential North-South

4) Nucleomagnetics with Neutron Separate of Proton Repulsion ('strong interaction' or gluon)

5) Nucleomagnetics Repulsion of Electrons ('weak force')

6) Net Electrostatic Force integral over time after quantized Nucleomagnetics separation ('gravity') of Positive electrostatic charge protons from Negative electrostatic charge electrons

Endnotes

ⁱⁱ Picture courtesy of NASA.

ⁱⁱ Using Coulomb's Law, the Charge-Force at 1-meter is:

$$F = k_e \frac{q_1 q_2}{d^2}$$

Factor	Net-Charge Calculation
Charge Force factor	k = 10^{10} m² / (s²) [9.03×10⁺⁹]
Charge of orbiting Electron	Q=10^{-19} [1.602×10⁻¹⁹]
Charge of distance Proton	Q=10^{-19} [1.602×10⁻¹⁹]
Distance	d=1 m or 10^0
Exponent shortcut	*+k+Q+Q-d-d*
Gross Charge Short-cut	*10-19-19-0-0 =*
calculation	*10-38-0 = -28*
Gross Charge Force	10^{-28} m¹ / (s²)

ⁱⁱⁱ

For the Advanced Explorer:

If particles actually have no dimensions, as some contend, there is challenge is at very close the 1/distance-cubed would become infinitely huge, but I disregard as the particle must have dimensions and thereby dimensions less than a particle size where that anything divided by zero (1÷0) = **infinity** would not and does not apply. Zero dimension charge particles (protons, electrons) do not exist.

Therefore, this proof assumes that particles do have physical dimensions (length, width, height).

[iv] https://en.wikipedia.org/wiki/Nuclear_force

[v] https://en.wikipedia.org/wiki/Proton

[vi] http://www.bing.com/search?q=distance+of+Bohr+radius&src=IE-SearchBox&FORM=IENTTR&conversationid=

[vii] Wikipedia

www.ingramcontent.com/pod-product-compliance
Lightning Source LLC
Chambersburg PA
CBHW040805200526
45159CB00022B/21